# Global Displacements

# Antipode Book Series

Series Editors: Vinay Gidwani, University of Minnesota, USA and Sharad Chari, CISA at the University of the Witwatersrand, South Africa

Like its parent journal, the Antipode Book Series reflects distinctive new developments in radical geography. It publishes books in a variety of formats – from reference books to works of broad explication to titles that develop and extend the scholarly research base – but the commitment is always the same: to contribute to the praxis of a new and more just society.

## Published

Global Displacements: The Making of Uneven Development in the Caribbean
*Marion Werner*

Banking Across Boundaries: Placing Finance in Capitalism
*Brett Christophers*

The Down-deep Delight of Democracy
*Mark Purcell*

Gramsci: Space, Nature, Politics
*Edited by Michael Ekers, Gillian Hart, Stefan Kipfer and Alex Loftus*

Places of Possibility: Property, Nature and Community Land Ownership
*A. Fiona D. Mackenzie*

The New Carbon Economy: Constitution, Governance and Contestation
*Edited by Peter Newell, Max Boykoff and Emily Boyd*

Capitalism and Conservation
*Edited by Dan Brockington and Rosaleen Duffy*

Spaces of Environmental Justice
*Edited by Ryan Holifield, Michael Porter and Gordon Walker*

The Point is to Change it: Geographies of Hope and Survival in an Age of Crisis
*Edited by Noel Castree, Paul Chatterton, Nik Heynen, Wendy Larner and Melissa W. Wright*

Privatization: Property and the Remaking of Nature-Society
*Edited by Becky Mansfield*

Practising Public Scholarship: Experiences and Possibilities Beyond the Academy
*Edited by Katharyne Mitchell*

Grounding Globalization: Labour in the Age of Insecurity
*Edward Webster, Rob Lambert and Andries Bezuidenhout*

Privatization: Property and the Remaking of Nature-Society Relations
*Edited by Becky Mansfield*

Decolonizing Development: Colonial Power and the Maya
*Joel Wainwright*

Cities of Whiteness
*Wendy S. Shaw*

Neoliberalization: States, Networks, Peoples
*Edited by Kim England and Kevin Ward*

The Dirty Work of Neoliberalism: Cleaners in the Global Economy
*Edited by Luis L. M. Aguiar and Andrew Herod*

David Harvey: A Critical Reader
*Edited by Noel Castree and Derek Gregory*

Working the Spaces of Neoliberalism: Activism, Professionalisation and Incorporation
*Edited by Nina Laurie and Liz Bondi*

Threads of Labour: Garment Industry Supply Chains from the Workers' Perspective
*Edited by Angela Hale and Jane Wills*

Life's Work: Geographies of Social Reproduction
*Edited by Katharyne Mitchell, Sallie A. Marston and Cindi Katz*

Redundant Masculinities? Employment Change and White Working Class Youth
*Linda McDowell*

Spaces of Neoliberalism
*Edited by Neil Brenner and Nik Theodore*

Space, Place and the New Labour Internationalism
*Edited by Peter Waterman and Jane Wills*

## Forthcoming

Money and Finance After the Crisis: Critical Thinking for Uncertain Times
*Edited by Brett Christophers, Andrew Leyshon and Geoff Mann*

Fat Bodies, Fat Spaces: Critical Geographies of Obesity
*Rachel Colls and Bethan Evans*

Enterprising Nature: Economics, Markets and Finance in Global Biodiversity Politics
*Jessica Dempsey*

The Metacolonial State: Pakistan, Critical Ontology, and the Biopolitical Horizons of Political Islam
*Najeeb A. Jan*

Taken For A Ride: Neoliberalism, Informal Labour And Public Transport In An African Metropolis
*Matteo Rizzo*

The Impunity Machine: Genocide and Justice in Guatemala
*Amy Ross and Liz Oglesby*

Frontier Road: Power, History, and the Everyday State in the Colombian Amazon
*Simón Uribe*

# Global Displacements

*The Making of Uneven Development in the Caribbean*

Marion Werner

**WILEY** Blackwell

This edition first published 2016
© 2016 John Wiley & Sons, Ltd.

*Registered Office*
John Wiley & Sons, Ltd., The Atrium, Southern Gate, Chichester, West Sussex,
PO19 8SQ, UK

*Editorial Offices*
350 Main Street, Malden, MA 02148-5020, USA
9600 Garsington Road, Oxford, OX4 2DQ, UK
The Atrium, Southern Gate, Chichester, West Sussex, PO19 8SQ, UK

For details of our global editorial offices, for customer services, and for information about
how to apply for permission to reuse the copyright material in this book please see our
website at www.wiley.com/wiley-blackwell.

The right of Marion Werner to be identified as the author of this work has been asserted in
accordance with the UK Copyright, Designs and Patents Act 1988.

*Library of Congress Cataloging-in-Publication Data*

Werner, Marion.
   Global displacements : the making of uneven development in the Caribbean / Marion Werner.
      pages   cm. – (Antipode book series)
   Includes bibliographical references and index.
      ISBN 978-1-118-94199-7 (hbk.) – ISBN 978-1-118-94198-0 (pbk.)   1. Clothing trade–
Caribbean Area.   2. Caribbean Area–Economic conditions–Regional disparities.
3. Globalization–Caribbean Area.   I. Title.
   HD9940.C272 W47 2015
   338.9729–dc23
                                                      2015022476

A catalogue record for this book is available from the British Library.

Cover image: Food vendors outside a garment factory, Santiago trade zone, Dominican Republic
© Marion Werner, with Andy Lu

Set in 10.5/12.5pt Sabon by SPi Global, Pondicherry, India

Printed in the UK

# Contents

# Series Editors' Preface

The *Antipode Book Series* explores radical geography "antipodally," in opposition, from various margins, limits, or borderlands.

*Antipode* books provide insight "from elsewhere," across boundaries rarely transgressed, with internationalist ambition and located insight; they diagnose grounded critique emerging from particular contradictory social relations in order to sharpen the stakes and broaden public awareness. An *Antipode* book might revise scholarly debates by pushing at disciplinary boundaries, or by showing what happens to a problem as it moves or changes. It might investigate entanglements of power and struggle in particular sites, but with lessons that travel with surprising echoes elsewhere.

*Antipode* books will be theoretically bold and empirically rich, written in lively, accessible prose that does not sacrifice clarity at the altar of sophistication. We seek books from within and beyond the discipline of geography that deploy geographical critique in order to understand and transform our fractured world.

Vinay Gidwani
*University of Minnesota, USA*

Sharad Chari
*CISA at the University of the Witwatersrand, South Africa*

**Antipode Book Series Editors**

# List of Abbreviations

| | |
|---|---|
| ADIH | Association des Industries d'Haïti (Haitian Industry Association) |
| APEDI | Asociación para el desarollo, Inc. (Association for Development, Inc.) |
| CARICOM | Caribbean Community |
| CBERA | Caribbean Basin Economic Recovery Act |
| CBI | Caribbean Basin Initiative |
| CBTPA | Caribbean Basin Trade Partnership Act |
| CD | Convergence Démocratique (Democratic Convergence) |
| CMT | Cut-make-trim |
| CNZFE | Consejo nacional de zonas francas de exportación (National Trade Zone Council) |
| CODEVI | Compagnie de Développement Industriel (Industrial Development Company) |
| DR-CAFTA | Dominican Republic-Central America Free Trade Agreement |
| DSNCRP | Document de Stratégie Nationale pour la Croissance et la Réduction de la Pauvreté (National Strategy for Growth and Poverty Reduction) |
| FEDOTRAZONAS | Federación dominicana de trabajadores de zonas francas (Dominican Federation of Trade Zone Workers) |
| FL | Fanmi Lavalas (Lavalas Family) |
| FTZ | Free trade zone |
| GATT | General Agreement on Tariffs and Trade |
| GDP | Gross Domestic Product |
| GSP | General System of Preferences |
| GOH | Government of Haiti |

| | |
|---|---|
| HOPE | Haitian Hemispheric Opportunity through Partnership Encouragement Act |
| IDB | Inter-American Development Bank |
| IFC | International Finance Corporation |
| IFI | International financial institution |
| IHRC | Interim Haiti Recovery Commission |
| ILO | International Labor Organization |
| IMF | International Monetary Fund |
| ISI | Import substitution industrialization |
| ITUC | International Trade Union Confederation |
| MFA | Multifibre Arrangement |
| MINUSTAH | United Nations Stabilization Mission in Haiti |
| NAFTA | North American Free Trade Agreement |
| OAS | Organization of American States |
| OPT | Outward processing trade |
| PARDN | Plan d'Action pour le Relèvement et le Développement National d'Haïti (Action Plan for National Recovery and Development) |
| PNUD | Programa de las Naciones Unidas para el desarollo (United Nations Development Programme) |
| PRSP | Poverty Reduction Strategy Paper |
| SOKOWA | Sendika Ouvriye Codevi Wanament (Union of Codevi Workers, Ouanaminthe) |
| UN | United Nations |
| USAID | United States Agency for International Development |
| WTO | World Trade Organization |

# List of Figures and Tables

## Figures

## Tables

# Acknowledgements

I owe a great debt to those who started me on this path and accompanied my journey long before I ever thought of writing a book: Altha Cravey, Tyrell Haberkorn, Leon Fink, Maria Mejía Perez, Monica Felipe Alvarez, Lynda Yanz, and Jennifer Hill, as well as the writings of Chandra Mohanty and Cynthia Enloe.

The research for this book began in 2004. I am grateful to the University of Minnesota's Department of Geography, the Graduate School, and the Interdisciplinary Center for the Study of Global Change (formerly the Macarthur Program) for generously funding my fieldwork and graduate studies. The University of Minnesota offered a vibrant intellectual community. I am deeply appreciative of my teachers and mentors who generously gave of their time and offered many insights as this project developed: Vinay Gidwani, Kale Fajardo, Abdi Samatar, Michael Goldman, and the community of graduate students in the Department of Geography and the Macarthur Program. I am especially grateful to Eric Sheppard and Richa Nagar who guided this project at its early stages and who have become role models and friends over the past decade as I entered the academy as a faculty member. Richa's indefatigable commitment to knowledge production that is accountable to working women and men in the global South has been a constant source of inspiration to me. I owe much to Eric's work on uneven development as well as to his singular style of accompanying an intellectual process: at once rigorous, respectful, and unbelievably patient.

Numerous individuals and institutions made my work in the Dominican Republic possible. Robin Derby connected me to local academics and persistently supported me to pursue a research agenda in the country. José-Leopoldo Artiles provided invaluable intellectual encouragement, insights, and contacts. I am also grateful to Pavel Isa, Miguel Ceada-Hatton, Lourdes Pantaleón, Leopoldo Valverde, Haroldo

Dilla, and Bridgette Wooding. Andrew Schrank generously offered initial background information and contacts with the trade zone sector. I am grateful to the executive director and staff of the Association of Trade Zone Industries of Santiago who facilitated my research and my ability to gain access to firms. This book owes a great debt to the rank and file members and former workers of IA Manufacturing who shared their stories with me, and to the local organizers of FEDOTRAZONAS for helping me to connect with them. Since 2009, I have had the fortune of knowing Marcos Morales of the School of Geography at the Autonomous University of Santo Domingo. I am grateful to Marcos and the School for their continued hospitality and for enriching dialogue.

Again, Robin Derby offered invaluable advice and encouragement to pursue my questions to the border with Haiti and beyond. My study of the border trade zone and local politics in Ouanaminthe would not have been possible without the insights and support of Yannick Etienne, Delien Blaise, and Lorraine Clewer. I traveled to Haiti again in 2012 to conduct interviews on post-earthquake reconstruction. My research was made possible by the assistance of two extremely talented students while in Port-au-Prince, Nicolas Pradhally and Carson Lagrandeur. I am also grateful to Jane Regan and Haiti Grassroots Watch.

I wrote parts of this book during a post-doctoral fellowship funded by the Canadian Social Sciences and Humanities Research Council. During my time in Toronto, I benefited from the wisdom and support of Leah Vosko, Rachel Silvey, and Katharine Rankin. During this period, I also met Gavin Smith and Winnie Lem, who have provided an academic home away from home on numerous subsequent visits to Toronto, for which I am extremely grateful.

This project took shape as a book over the past four years while I was at SUNY-Buffalo (UB). I am grateful to supportive colleagues in the Department of Geography, especially Trina Hamilton, and to friends and colleagues around campus. I am indebted to UB's Humanities Institute for a fellowship that permitted me to complete the manuscript. I also want to thank UB's Gender Institute for hosting a research workshop on a portion of the manuscript. Marissa Bell offered incredible assistance in preparing references and Andrew Lu did wonderful work adapting the cover image for publication.

Several colleagues and friends encouraged me to pursue this project and generously offered comments, critiques, and advice at various stages including Arturo Victoriano, Bradley Wilson, Joel Wainwright, Melissa Wright, and Beverley Mullings. Yasmine Shamsie and Alex Dupuy both contributed helpful comments on Chapter 6. Josh Barkan read this project at its various stages. I am indebted to his keen analysis of the problematic of the book – in the sense of this book's theoretical terrain *and* the university as a site of knowledge production – without which I am not sure this book would exist. The generous and insightful comments of two anonymous reviewers, I believe, have significantly improved the manuscript. Many of the concerns at the heart of this book were nursed and formed into intellectual arguments through my long-standing dialogue with Jennifer Bair. I am deeply indebted to Jenn for her unflinching critique and intellectual generosity over the past six years. All errors of interpretation are of course my own.

From manuscript to commodity: I want to acknowledge the support of Vinay Gidwani and Sharad Chari, as well as Jacqueline Scott at Wiley Blackwell, who guided me patiently through the process. The commodity chain extends further to the (transnational) labor of copy editors, type-setters, press operators, bindery feeders, and transportation workers, whose hidden work I gratefully acknowledge. I also thank Hannah Lilien for her incredible care of my infant son during the last stages of this project.

Finally, I extend my deepest gratitude to close friends and family: Cynthia Tan, Hilary Masemann, Alison Chan, May Friedman, Ron Vida, Heather Walters, Jack Gieseking, Sara Koopman, Diana Ojeda, Libby Lunstrum, Anna Falicov, and Anyelina Fernandez, my brothers, Martin and David, and their wonderful families, and especially my parents, Beatrice and Daniel, who have been a source of constant love and support. Thank you. Jaume Franquesa not only supported this project intellectually and emotionally, but also afforded me innumerable mornings and evenings of extra time to finish the manuscript. His encouragement, love, patience, and care have made all the difference. And finally, I am grateful for life's treasures, Yannick and Biel.

Buffalo, New York
April 11, 2015

# 1

# Introduction

## *Power and Difference in Global Production*

*The violence of abstraction produces all kinds of fetishes: states, race, normative views of how people fit into and make places in the world. A geographical imperative lies at the heart of every struggle for social justice; if justice is embodied, it is then therefore always spatial, which is to say, part of a process of making a place.*
— Ruth Wilson Gilmore, "Fatal Couplings of Power and Difference," p. 16

*This is a geography, not of jobs but of power relations, of dominance and subordination, of enablement and influence, and of symbols and signification.*
— Doreen Massey, *Spatial Divisions of Labour*, p. 3

## Sundays in Los Almácigos, Dominican Republic

In the late afternoon on a Saturday in May, Ambrosina and I met up at the bus station in the city of Santiago, the inland industrial capital of the Dominican Republic's northern region, called the Cibao. We were headed back to her hometown of Villa de Los Almácigos, west of the city, about an hour from the Haitian border. Ambrosina lost her job in February along with 6,000 other trade zone workers when the garment factory where she worked, IA Manufacturing, closed down. She agreed to help me interview her former co-workers in her hometown.

*Global Displacements: The Making of Uneven Development in the Caribbean*, First Edition.
Marion Werner.
© 2016 John Wiley & Sons, Ltd. Published 2016 by John Wiley & Sons, Ltd.

On the bus ride, Ambrosina told me about her current job. After two months of pounding the pavement and handing out dozens of résumés, she had gone to work for her former supervisor as a final auditor at the country's largest surviving garment firm, Dominican Textile. This would be her fourth garment company since finishing high school and moving to Santiago seven years earlier to start university. For the next two hours, as we traveled through the verdant countryside, Ambrosina explained in detail the many things she disliked about her new job: more work, less pay, submissive employees, and frantic managers.

The previous week, her manager had increased her module's quota for Old Navy shirts from 5,000 to 6,000, offering the workers 200 pesos – a little more than 7 dollars – for the new weekly production goal. Ambrosina was galled by the operators' acceptance of such a measly bonus. Despite being proud of her rural roots, she attributed her co-workers' acquiescence to the fact that many were from the *campo* (rural area) where the factory was located just outside the city. She also resented the long commute to her new job, adding an extra two hours onto her already grueling 10- to 12-hour work shifts.

*When I think of the things we used to complain about at IA*, she lamented, *it seems unreal*. Ambrosina would repeat her complaints about her new job to family members and former co-workers whom we would visit the following day in Los Almácigos. They would shake their heads and chime in periodically with sympathetic exclamations, "Jesús!" "Muchacha!" She would quit a few months later.

The next morning, we started on foot down the two-lane, paved road that eventually links Los Almácigos to the border with Haiti, 60 kilometers away. Ambrosina was somewhat mortified to be seen walking in town but I hoped she would just blame it on me. Our time was tight and I felt that we could ill afford to wait for her father to return with his motorbike so that we could use it for travel to the houses nearby. As on our other weekend visits, our interviews were squeezed between morning house cleaning, lunch preparation, and other domestic chores at her parents' house, intensified by the fact that Sunday was the only day with some consistent electricity supply, and thus also the best day to do laundry and ironing.

After three brief visits to the houses of former co-workers, we arrived at Leidy's family's home. Leidy had returned to Los Almácigos just two weeks prior. She had worked in three different factories in the trade zone over six years. Her longest stint was at IA Manufacturing, where she sewed front pockets onto Dockers-brand pants. After the factory closed, she worked in a five-machine workshop sewing shorts and pants for the domestic market in one of Santiago's working class neighborhoods. She learned new operations, but didn't always get paid regularly and decided she couldn't risk more of her time working for free.

Leidy's mother joined our conversation. She and her husband, Leidy's father, had gone to the trade zone in Santiago with Leidy and three of her siblings in 1992, closing up their small house in Los Almácigos, and leaving behind the mother's plot of land. Her parents worked in a garment factory for a little more than a year. *Despite the problems today,* her mother said, *conditions were worse and wages were lower back then. We returned with nothing,* she concluded bitterly. At the age of 18, nearly a decade after her parents' failed attempt, Leidy told the family she was going to the trade zone. Her parents warned her against the move, but she went anyway as she was determined to find her independence. Now, back at her parents' place with her young daughter, Leidy was making arrangements to bring her belongings from Santiago. There was a small, one-room wood house behind her parents' modest *rancho* that she would make her home. *I'm not going back,* she told us. Although seemingly resigned, Leidy said she was waiting to hear from her brother-in-law about a possible modeling opportunity in the city. Ambrosina consoled Leidy by sharing her complaints about her new job at Dominican Textile. As we were leaving, Ambrosina offered to help Leidy obtain a social security benefit from the government for her daughter who had been born while she was still an employee (and thus she was entitled to a small monthly payment). *If only there were jobs here, there'd be no reason to have to go to Santiago,* Ambrosina told her as we departed. I wasn't so sure Leidy shared Ambrosina's love for her *campo,* but for now she would have little choice but to make her life there.

## Situating Global Displacements

Why did Leidy not heed her parents' warnings? What expectations of progress motivated the migrants from Los Almácigos and other precariously employed trade zone workers? And what is at stake in situating a study of contemporary globalization that takes the threshold of these workers' idled garment factory as its point of departure? These questions are the fertile soil from which the present study has sprung.

The experiences of former garment workers in Los Almácigos do not square easily with conventional accounts of the globalization of production. These accounts generally begin with the observation of a fundamental reorganization in the geography of production starting in the late 1960s. The substantial surge in manufacturing output outside of "core" capitalist countries prompted a broad swath of observers to proclaim the dawn of a new structural relationship between what was then commonly called the First and Third Worlds (Fröbel et al. 1980; Piore & Sabel 1984; Lipietz 1987; Harvey 1989). Up until the late 1970s,

Third World states forged this reorganization of industry primarily through the promotion of strategic industrial sectors and the protection of domestic markets, while multinational corporations adapted to and, in turn, shaped these geographies through foreign direct investment. The demise of this import substitution industrialization (ISI) strategy by the end of that pivotal decade dramatically transformed the configuration and organizational form of the global division of labor. As the United States sought new conditions to maintain its dominance in the face of the limits of the post-World War II economic arrangement (Arrighi 1994; Krippner 2011), much of the Third World faced insolvency brought on by high interest rates, capital scarcity, higher oil prices, declining resource prices, and a weakened industrial sector (McAfee 1991; Corbridge 1993).

By the end of the 1980s, Latin America's primarily US creditors, together with multilateral development banks, implemented a series of policies that privileged so-called export-oriented industrialization, or the production of labor-intensive manufactured goods for Northern markets. This model, already in operation in a handful of experimental zones in the region such as the United States–Mexico border and the Haitian capital of Port-au-Prince, would soon become dominant as states progressively re-regulated trade and investment in line with multinational corporations' priorities. This turn towards what is commonly called neoliberalism was further entrenched in new generation trade agreements, beginning with the North American Free Trade Agreement (NAFTA), which sought to codify the priorities of multinationals and finance capital in global trade (Cox 2008).

The "global factory" – manufacturing facilities in Latin America, Asia, and Africa making the innumerable products filling seemingly endless store shelves in North America and Europe – has since become iconic of this reorganization of transnational production, and of globalization in general. Feminist ethnographers were quick to elucidate the gender dynamics of this new model, which was not simply export-led but also female-led (Standing 1999). From Indonesia and Malaysia to Mexico and Haiti, the workers in these factories were overwhelmingly women. In East and Southeast Asia, these young "factory daughters" (Wolf 1992) faced an intensification of household duties while also experiencing newfound forms of limited independence (e.g., Lee 1998; Ngai 2005; Ong 1987). In Latin America and the Caribbean, women in export factories were often taking on the "breadwinner" role under dramatically eroded wage conditions as their male kin faced the indignities of decline in import substituting industrial sectors (e.g., Elson & Pearson 1981; Fernández-Kelly 1983; Cravey 1998; Mullings 1999; Salzinger 2003; Wright 2006; see Bair 2010 and Cravey 2005 for reviews and discussion).

The present study is deeply indebted to this diverse body of feminist work and draws on its insights to extend feminist analysis to a period of major restructuring following more than 30 years of the low-wage export model. In the extensive body of English-language literature on the global factory, feminist and otherwise, scholars have tended to follow capitalist investment and contracting relationships to "far-flung" locations (from the perspective of metropolitan scholarship) and have theorized the resulting new divisions of labor therein. Far fewer, however, have paid attention to the transformation of these factory arrangements as initial low-wage production has given way to new combinations of technology and skill, together with mass retrenchment and disinvestment.[1] The aim of this book is to study this contemporary process of restructuring of the global factory, and in so doing, to inform thinking and action on the persistent and dynamic relations of inequality that articulate the global North and the global South.[2]

One obstacle to understanding the mutable yet structured global division of labor is the way that it is mapped onto sequentialist understandings of change and fetishized geographical categories. Consider, for example, the common "three worlds" classification of developing, newly industrializing, and post-industrial countries. As Fernando Coronil (1996) argued, such nomenclatures obscure the politics of these geohistorical categories: the ways in which places and their positioning in the global economy are forged through *connections* rather than distinct and sequential *transitions*. As we will see in the chapters that follow, while scholars, development experts, and activists dispute the merits of the global factory, few dispute the notion that these installations will either push, or fail to push, places along a path from farm to factory to the post-industrial. The global factory, then, is not only a relationship of subcontracting, outsourcing, investment, or exploitation; it is *also* a set of assumptions, discourses, and spatial imaginaries that reproduce the notion of development as one of traversing through a stage of industrialization.[3]

The notion of a stage-like trajectory of capitalist development is inseparable from the reproduction of Eurocentric dualisms of West/non-West and developed/developing. I follow Anibal Quijano's notion of Eurocentrism as "the paradoxical amalgam of evolution and dualism" that reproduces Europe and its settler colony extensions, principally the United States, as a self-actualizing center, which defines the Other in relation to the Self (2000a: 551). In exploring the links between Eurocentric dualisms and the global factory, my intention is neither to assert a particular "Caribbean" path of capitalist development nor to claim the failure of "Western" development in the Caribbean. Rather, my aim is to show how the operation of global factories is inseparable from a discourse that frames global production as a stage of development. In so doing, this discourse produces normative assumptions of how racialized and gendered subjects, and the places where they live and work, fit into the global division of labor (see Gilmore, epigraph). This discourse does not simply *describe* the world; rather, it seeks to *transform* it by reworking multiple livelihood strategies, colonial legacies, and discontinuous industrial trajectories into dual economy models, narratives of transition, and, ultimately, newfound objects of capitalist development.

The critique of the development discourse that I offer in the following chapters is a contribution to, but not a substitute for, a political project to reframe and forge global connections differently. Towards this end, *Global Displacements* offers the reader a set of conceptual tools and analytical insights to engage with the complex geography of industrial and deindustrial processes that find temporary stability in given arrangements of factories, workers, and consumers. I develop these tools by recuperating the notion of uneven development from the archives of Marxism towards the construction of a critical, feminist approach to the study of global production.

The concept of uneven development stems from early twentieth-century debates over the possibilities of socialist revolution in countries that remained primarily agrarian (Trotsky 1969 [1906]; see also Burawoy 1989). In the immediate post-World War II period, the concept was incorporated into theories of structural dependency between the Third and the First Worlds (e.g., Frank 1967; Amin 1976). In the throes of deindustrialization in the North Atlantic in the 1970s and 1980s, Marxist geographers provided a spatially nuanced understanding that eschewed taken-for-granted scales of uneven development as either a relation among nations or between world regions. Instead, their work showed how the ordinary process of capitalist accumulation reproduced spatial unevenness. Capital sought to resolve contradictions between fixity and mobility, competition and monopoly, and exploitation and consumption through the production of novel spatial arrangements and scales (Massey 1995 [1984]; Harvey 1999 [1982]; Smith 2008 [1984]).

In the present text, I develop a formulation of uneven development through an understanding of global production arrangements as concrete determinations of social and spatial divisions of labor. The concept foregrounds relations of power – among capitals and between capital and labor – *and difference* – forged through colonial legacies and everyday politics – that are obscured by path-like notions of industrial change. I engage with Marxist and feminist approaches shaped and reinterpreted through my ethnographic work on production restructuring in the export garment industry in the Dominican Republic and Haiti.

By confronting the messiness of daily life, the irreducibility of experience, and the complexity of difference, I consider the transformation of this accumulation strategy woven through these two deeply inter-related countries. My grounded and theoretically informed account of employment collapse and restructuring in the Dominican Republic, combined with a new factory boom in Haiti, reveals the historically patterned and contingent production of new geographies of uneven development in the Caribbean. What emerges is an understanding of uneven development as a process that is *made* through the production of place and social difference as workers and the unemployed forge their livelihood possibilities under conditions not of their own choosing. Before undertaking a more extended discussion of uneven development, I turn first to an explanation of why this approach has emerged from my study of the Caribbean.

## From a Comparative to a Relational Geography of Haiti and the Dominican Republic

Perhaps nowhere in Latin America have emerging geographies of uneven development become more evident than on the island of Hispaniola. In the late 1960s, the Dominican Republic and Haiti had nearly the same gross domestic product (GDP), but by the 2000s, Dominican GDP was *eight times* that of Haiti.[4] Indeed, the Dominican Republic has enjoyed the second highest growth rate in Latin America since the turn to neoliberal regulation, a fact that is drawn upon by mainstream economists like John Williamson to extol the so-called Washington Consensus, or the structural adjustment policies that swept the region in the 1980s (Kuczynski & Williamson 2003). Haiti, on the other hand, has faced deteriorating economic and social conditions, a protracted political crisis and, as a result, profound vulnerability to socionatural disasters. In contrast to the celebratory narrative of neoliberal reform in the Dominican case, few mainstream economists attribute Haiti's dismal trajectory to the very similar, in fact even more radical, neoliberal policies adopted by the country in the 1980s and 1990s.[5]

These divergent trajectories have brought these two countries to broader public consciousness. In the popular book *Collapse* (2005), for example, author Jared Diamond dedicated a chapter to the profound disjuncture between Haiti and the Dominican Republic in Malthusian terms: Haiti is a place that suffers from too many people and not enough land, he argued, with deforestation as the starkest marker of land pressures. Taking the island – and its two nations – as akin to a natural experiment involving two separate societies "growing" under similar conditions, Diamond ultimately attributed Dominican "success" to a more reasonable elite and cultural difference.[6]

Such comparisons between Haiti and the Dominican Republic produce sequentialist and Eurocentric framings of development *par excellence*. The Dominican Republic apparently has progressed along a natural path from an agrarian to an industrial to a service-oriented (or post-industrial) economy, while Haiti has apparently failed to traverse these stages of development. These are powerful motifs of linear and territorially discrete social change that inform the governing logics and discourses of development policy. The always becoming of capital is annexed to dual economy models and transition narratives, premised on a political amnesia. As a result, we will see that the global factory is presented as a development imperative for Haiti, and as a stage that has passed for the Dominican Republic.

The power of such development narratives, and the comparisons that they engender, lies as much in what they foreground as in what they render invisible. In this case, to begin with, such a framing erases the distinct incorporation of what we now call Haiti and the Dominican Republic into colonial capitalism, as colonial Saint-Domingue, a slave plantation society and the jewel of the French empire, and as Santo Domingo, a mixed-race, impoverished frontier society, respectively (Derby 1994). Accounts like Diamond's also minimize the significance of imperial legacies and neo-colonial occupations: for example, in the case of Haiti, a crippling indemnity paid to France as a condition for receiving metropolitan recognition.[7] Such comparisons mask the mass migrations from both sides of the border that have effectively created transnational nations, on the one hand, and states highly dependent upon migrant remittances, like the rest of the Caribbean, on the other. This transnationalism has been produced by migrant flows to the global North since the 1960s, where between one-ninth and one-eighth of Dominicans and Haitians, respectively, now live.[8] There is another dimension to this transnationalism, however: the nearly century-long migratory flows from Haiti to the Dominican Republic, instituted by the US occupation of the island. This intra-insular integration has constituted and reproduced racialized and segmented labor markets, the implications of which we will examine in the chapters to come.

In short, the comparative framework relies on the notion of these two countries as territorial containers and interprets their differences as evidence of national success or failure in a world of sequential development. In contrast, my argument in the pages that follow is for a *relational geography* of capitalist uneven development that foregrounds the ways in which places are iteratively forged in relation to one another. For if the Dominican Republic and Haiti have long been complex, heterogeneous, "modern"[9] and unevenly integrated societies, it is since the latter part of the twentieth century, with the rise of neoliberal globalization, that their divergence has increasingly become a basis for capital accumulation through the restructuring of global production. Before sharing with the reader how I undertook a grounded study of this process of restructuring in and through Haiti and the Dominican Republic, I introduce the reader to the theoretical tools that I have drawn upon to develop the notion of the uneven development that I use in this text.

## Theorizing Uneven Development: Global Production Networks, Coloniality, and the Production of Place

To interpret the analytical significance of these divides in the Caribbean, I draw upon a mix of theoretical tools that I briefly sketch out here. Uneven development is a classic approach to understanding the material ways that places relate to one another through transfers of "surplus" from periphery to center. Throughout much of the twentieth century, scholars of Latin American development debated the forms of dependency that reproduced uneven development between Euro-American metropoles and the Third World. So-called *dependentistas* assigned primacy to these imperial-type relations among nominally sovereign states and attributed the persistence of global income inequality to them.[10] Their work was influential in the formulation of world-systems theory in Anglo-American scholarship, which argued for an understanding of capitalism at the world scale formed through exchange relations between a core, a semi-periphery, and a periphery. World-systems scholars Hopkins and Wallerstein (1977; 1986) introduced the concept of the commodity chain to disrupt predominant understandings of economic development as a national and sequential process by foregrounding the linked activities – from inputs to production, distribution and consumption – that produce a commodity.[11] Through unequal exchange along the commodity chain, they argued that hierarchy in the world system was reproduced. Core, semi-periphery, and periphery positions were not fixed, however. A certain mobility between positions existed as states and capitals collaborated and competed to shift from concentrating the production of lower value-added goods to higher value-added ones, or suffered reversals in such efforts (Arrighi & Drangel 1986).

Since its introduction by world-systems scholars, sociologists developed the idea of the *global* commodity chain as a meso-level framework to analyze how functionally integrated but spatially dispersed production activities shape development trajectories (Gereffi & Korzeniewicz 1994). Analysis of global commodity chains, now more commonly called global production networks, has facilitated scholars, as well as activists and policymakers, to gain analytical purchase on the complex and concrete determinations of the global economy (see Bair 2005; 2009).[12] The emergence of this analytical perspective parallels and documents a substantive shift in the organization of production over the past 40 years. Since the 1970s, multinational corporations either systematically disinvested from production, or increasingly made more profits from their financial and non-productive activities (Milberg 2008). In turn, these organizations shifted in form from vertical integration and direct control of production to relations of exchange and arms-length coordination and control between formally independent units (Piore & Sabel 1984; Harvey 1989; Arrighi 1994).

The study of the firm networks spanning the global North and South that have resulted from this shift from "fixity" to "flexibility" has yielded important insights into novel forms of unevenness produced through these relationships. Drawing on the work of Joseph Schumpeter (2008 [1942]), scholars have theorized the existence of higher value nodes in production networks, defined as activities that offer entrepreneurial "rents" – or above-average profits – through innovation and/or monopolization by a small number of firms (Kaplinsky 2000). In concrete terms, retailers and brands, based largely in the global North, occupy these more concentrated and monopolized nodes of the network, which offer greater returns. This "lead firm" position is reinforced by these firms' ability both to control the terms of participation of suppliers by setting prices and standards and to pocket the bulk of profits. In contrast to lead firms, supplier firms – some of which have consolidated into multinational companies themselves based largely in East Asia – generate their profits primarily through the increasingly competitive activity of production. Over time, as we will see in the case of the principal garment firm in the Dominican Republic, these producers also seek to enter niches of the chain where rents are possible. Firm strategies to create or access higher value activities vary between production networks. They can involve, for example, using a higher ratio of capital to labor, which usually also corresponds to a shift in the mix of activities that a given firm undertakes.[13]

The most useful insight from global production network studies for theorizing uneven development is the observation that different returns created and distributed via these networks produce geographic concentrations of wealth and poverty at a variety of scales (Arrighi & Drangel 1986).[14] Each node of a production network is not a set of "firms" but rather a given economic activity, or set of activities, that represents a mix of capital, labor, and land (Arrighi & Drangel 1986: 16). How returns generated by participation in production networks are created and distributed can only be understood through an analysis of relations of production (i.e., between capital and labor), together with those of exchange between productive units (i.e., firms, farms, households). Moreover, states play a central role in trying to attract capital and stabilize the conditions for investment (Arrighi & Drangel 1986. 24; see also Glassman 2011). In short, hierarchies in the global economy are determined neither by participation in particular sectors (i.e., agriculture versus industry), nor by fixed geographical locations (i.e., G-8 countries versus the rest of the world). Rather, uneven development is shaped by the different returns accrued to the mix of spatially situated "core-like" and "periphery-like" economic activities or functions. These value characteristics of economic activities are not static, moreover. They stem from the techno-organizational strategies of capital, supported by states, to hold highly competitive functions – that is, those with razor-thin margins – at arm's length through subcontracting and outsourcing, while capturing the monopolized, core-like functions that offer above-average returns.

As we think through the actual geographies of global production in light of these insights, one truism emerges: while the positions of places are not pre-ordained, a hierarchical *field* of possible positions exists. *By definition*, not all places can form part of the core at the same time. Uneven development, or the reproduction of places that concentrate core-like, periphery-like, or a mix of these activities, together with places that are excluded entirely from such arrangements, is thus the necessary architecture of a global economy dominated by capitalist production. In fact, as even staunch critics of world-systems theory have acknowledged, this unevenness has a certain regulatory function in the global economy (Lipietz 1986). The immanent contradictions of capitalism, between competition and monopoly, for example, "find" their core, periphery, and semi-periphery arrangements for a time. To reproduce the global economy, then, different positionalities among places must also be reproduced.

If a production network perspective has much to commend it in foregrounding how these relationships reproduce uneven geographies of development, the approach is limited in several key ways. Let me focus here on the two limitations that are most relevant to the present study: the question of labor and the politics of place.

First, the focus on competition between capitals (i.e., firms) linked through production networks generally leaves the question of labor under-theorized. In response, much scholarship concerned with the role of labor considers labor as either an object impacted by global production arrangements, or as an active agent shaping the form and content of these arrangements through collective struggles (see Bair & Werner 2015). I offer a third approach here that eschews an understanding of labor as equivalent to its role in a given production network. Instead, I consider labor through the lens of colonial legacies that intersect with particular transnational production arrangements but are not reducible to them. This approach is a first step towards conceiving of labor as an active structuring agent in the production of place.

The work of Anibal Quijano and his notion of the coloniality of power, or the historical and material (dis)continuities between colonialism and contemporary capitalism, is a helpful starting point. Quijano argues that the racial hierarchies forged through the conquest of the Americas constitute a terrain of articulation, one that adapts historic patterns of exploitation to the contingent necessities and attendant conflicts of contemporary capitalist accumulation. Capitalism is realized through the stitching together of the wage relation with other forms of labor control, structured by the value hierarchies of racialized and gendered labor. Contemporary arrangements of production reproduce – with difference – gendered and racial hierarchies of labor rooted in colonial legacies and exerting a structuring effect on the global division of labor of our times (Quijano 1998; 2000a; 2000b). As Sylvia Federici has argued, what Marx described as primitive, or originary, accumulation, was not solely the instantiation of capitalist relations of production through the separation of peasants, crafts- and tradespeople from the means of production, but also the introduction of social divisions of labor through the production of hierarchies of domination:

> Primitive accumulation ... was not simply an accumulation and concentration of exploitable workers and capital. It was *also an accumulation of differences and divisions within the working class,* whereby hierarchies built upon gender, as well as "race" and age, became constitutive of class rule and the formation of a modern proletariat. (Federici 2004: 63-64; emphasis in the original)

Federici offers an account of the devaluation of European women's labor in the process of the formation of capitalist relations of production in Europe, rendering women as devalued or value-less labor. In contrast, Quijano places greater emphasis on the originary relation between wages and European-ness/whiteness/maleness, and the correlating link between racialized and feminized subjects and unpaid work, forced labor, petty commodity production, and slavery.[15] For both, primitive accumulation is not a thing of the past but rather the on-going and uneven mix of forms of "extra-economic" violence and domination, together with the production of surplus value through labor exploitation, that create the conditions for capitalist accumulation and remake racialized and gendered hierarchies of labor.[16]

Quijano conceives of coloniality in the Americas as a "historical-structurally heterogeneous model of power with discontinuous relations and conflicts among its components" (2000a: 571). To put it simply, there is no straight line from the *encomienda* to the day laborers on Chiquita brand's contract banana plantations in Ecuador, or from slavery to the sewing operators in Levi's apparel suppliers in Haiti.[17] Moreover, Euro-centered racial categories and patriarchal gender formations constitute hierarchical systems of meaning, inextricably linked with, but irreducible to, capitalism. The notion of coloniality, then, does not presume any essential connection between subjects produced as Other through racialized and gendered material-symbolic hierarchies, and the actual relations of exploitation, domination, and exclusion of devalued subjects from market and wage relations. Neither capitalism, nor racism, nor patriarchy exists in a relation of necessary dependence. And yet, these structures are not simply analogical, as Spivak (1988) reminds us. Rather, there is a way in which accumulation proceeds through the iterative reproduction of coloniality: that is, through the reworking of hierarchies of social difference and forms of labor in order to recuperate profits from their interminable tendency towards stagnation and decline. In short, in global production networks, social difference serves as a resource to rework arrangements of exploitation, and to redraw the lines of inclusion and exclusion in the capitalist wage relation. To grasp this iterative reworking of coloniality, however, we must shift our analysis from the abstract lineaments of value and difference in global hierarchies of capital and labor to the concrete determinations and everyday politics of place production.

Places and regions are best considered as processes rather than pre-given objects for our analysis. As geographer Doreen Massey (1995 [1984]) argues, regions are themselves produced by complex, multi-layered social histories, and overdetermined by their integration in multiple space-times of accumulation. Their participation in (often) multiple circuits of capital accumulation is part of, but not reducible to, their on-going formulation (Massey 1995 [1984]). Massey's notion of place as process dovetails well with Gillian Hart's (1998; 2002) work on rural industrialization. Drawing on the Marxist agrarian tradition, Hart argues that rigid, binary approaches to industrial change fail to comprehend the specificity of place as the outcome of historically specific forms of property relations and on-going power struggles over how the social product will be produced and distributed. These "layered" histories of social relations, and on-going struggles to shape these relations, produce "multiple, nonlinear and divergent trajectories of capitalist development" (Hart 1998: 334). Moreover, these political economic struggles are inseparable from cultural ones. "[S]truggles over material resources, labor discipline and surplus appropriation [and distribution]," Hart writes, "are simultaneously struggles over culturally constructed meanings, definitions, and identities" (1998: 340).

In the Caribbean, the anthropological literature has long debated the multiple, divergent trajectories of regions incorporated into the global market via smallholder agriculture versus plantation production. This distinction was made famous by the comparison between social relations in the tobacco- and sugar-producing regions of Cuba in Fernando Ortiz's classic work *Cuban Counterpoint* (1995 [1947]). In this and much subsequent study on the plantation/peasant complex in the Caribbean, a theme I return to in Chapter 2, we see the inseparability of relations of production, global market integration, political struggle, and cultural production woven through particular regional formations (e.g., Mintz 1974; Hoetink 1980; Trouillot 1988). Such attention to regional trajectories of change disrupts binary understandings of uneven development as a process of incorporating peripheries into a totalizing and homogenizing "core" or "the West." In particular, a focus on the social relations that constitute regional formations in relation with – but irreducible to – foreign capital and markets disrupts the still pervasive understanding of global production as fundamentally linking low-wage production in the South to consumption in the North (Coronil 1996). As Priti Ramamurthy (2004) has argued in her call for a feminist commodity chain analysis, such binaries must be resisted for they fundamentally obscure the complex social processes in the global South that are constitutive of global capitalist relations.

The challenge that I take from Hart, Coronil, and Ramamurthy, together with the Caribbean and Latin American literature on regional change, is to specify the particular geographies of uneven development formed through colonial legacies, capitalist relations, and the ways that these forces produce regional conjunctures. In the chapters that follow, my aim is to convey an understanding of uneven development in and through global production networks that eschews a narrow economism and a one-dimensional conception of power. For if coloniality emphasizes on-going relations of domination and exploitation, I follow Stuart Hall in theorizing colonial legacies as "active structuring principles of the present" in particular historical and place-specific conjunctures (1980: 339).[18] Regions, then, are *not* slotted into global divisions of labor. Rather, regions are dynamically reproduced in relation to value hierarchies of production, *themselves not fixed* as I have discussed above. Regional formations must be interrogated as dynamic sets of relations that articulate firm hierarchies, heterogeneous racialized and gendered forms of labor, and political struggles over the form of capitalist accumulation and the distribution of the social product.

While I embrace Hart's (2002) notion of "multiple trajectories" in my analysis of export restructuring in Haiti and the Dominican Republic, my aim is to mobilize such an approach in relation to global production networks and the reworking of coloniality in the Caribbean. For if there are multiple trajectories of place formation, these paths are not plural. To deepen our understanding of global connections, then, my goal is to consider place production as a process that articulates with a shifting, heterogeneous, yet structured global economy. The reproduction of these multiple trajectories is itself a spatial resource of difference for the reproduction of uneven development and thus the global capitalist economy itself. If, as Melissa Wright (2006) has argued, capital requires "corporeal breadth" – that is, social differences of labor – to rework value hierarchies in production, capitalism also requires variegated places – *geographic breadth* if you will – to reproduce its relations. Far from fixed objects and positions, a feminist approach to uneven development calls for the continual interrogation of the dynamic basis of domination, exploitation, and creative forms of subject-making within and between places. An analysis of these historically patterned and contingent geographies offers us a way to understand global production networks as arrangements that exist in dialectical tension with regional trajectories of change, and the collective labor and subaltern struggles that shape and are shaped by these processes.

## Global Displacements as Multi-Sited Fieldwork

While *Global Displacements* is influenced by my engagement with these Marxist, feminist literatures, my feminist understanding of uneven development has emerged from my fieldwork with workers, owners, managers, and policymakers linked to the transnational garment industry and its operations in the Dominican Republic and Haiti.

I began this project wanting to expose the reality of factory closures and their effects on workers and communities as a way to disrupt teleological understandings of global production. I believed that an engaged study of a deindustrial process would simply reveal the mistaken assumptions behind the development discourse of global production. As I became more deeply involved with the project, however, and the forms of livelihood that circulated through different sites of production and social reproduction associated with the garment industry, the experiences that workers, managers, community leaders, and the un- and underemployed shared with me did not fit into such a singular perspective. To begin with, there is no underlying reality of exclusion and decline that is *masked by* industrialization and progress narratives. Rather, narratives of progress and uneven geographies of inclusion/exclusion are, I discovered, deeply imbricated with one another. What struck me immediately when I began interviewing workers and managers in the Dominican Republic, for example, was the pervasive understanding that the export garment model was part of a quickly retreating past for the country, echoing published development reports and industry analysis. Belief in this notion was underscored by garment production moving to Haiti, reinforcing a notion of path-like development despite the much more circuitous and interrupted routes that both countries have traversed. The current and former garment workers, managers, and owners who shared their experiences with me certainly did not all agree with this framing; nevertheless most constructed their thoughts and opinions about work and livelihood in relation to this dominant narrative. Thus, rather than a method that disrupted or disabled this hegemonic understanding of development by revealing a more authentic reality, my fieldwork dug deeply into everyday narratives and practices that pervaded the sector's restructuring. These accounts were rife with complex understandings of progress, ambivalent experiences of work, and the inextricable struggle over material livelihood and social positions of worth.

My fieldwork focused on three sociospatial locations: first, garment firms in Santiago enmeshed in local and international supply networks; second, the circuits of livelihood navigated by unemployed garment workers, especially those linking the city of Santiago to rural *campos* predominantly in the surrounding Cibao region; and finally, the historical and social conditions of place-making underlying the advent of industrial border production on the island in the town of Ouanaminthe (see Figure 1.1).

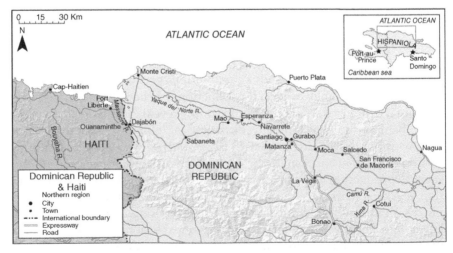

**Figure 1.1**   Map of the Dominican Republic and Haiti, northern region

To study garment firms, I spent several months based in the offices of the regional trade zone association. The association was located in a converted garment factory within Santiago's main trade zone, just off the busy *circunvalación* highway that circled this inland city. The director provided office space and opened innumerable doors for me to conduct initial interviews with firm owners and top executives of four firms, and general managers of 12 additional firms. The interviews covered basic company data and history, as well as the informant's assessment of the firm's competitiveness, the role of the state in promoting exports, and the future of the sector. Because the trade zone association largely represented Dominican garment firms, I spoke primarily to owners and managers of Dominican firms with the exception of a US-owned branch plant and a Hong Kong-based subsidiary, where managers granted me interviews as a favor to the association's director. Interviews were often followed by a factory tour. I returned to eight firms for more open-ended follow-up interviews. My questions developed over time as I incorporated the perspectives shared by owners and managers about the particular characteristics of the northern region, the Cibao, and its cultural attributes that secured the sector's competitiveness. Thus, I came to understand the location of these garment firms as one produced through their commercial relations to US apparel multinationals, as well as through a particular (subnational) regional culture and an associated trajectory of development.

Participant observation within firms was more restricted for reasons I explain below. In order to gain more unstructured access to the sector, I took a series of night classes with engineers, offered by a government- and US-funded Textile Center established to reskill a garment industry that hitherto had thrived on simple assembly. Effectively, the Textile Center was a classroom area with a handful of computers, a couple of dressmakers' dummies, and standing tables in the same factory building as the trade zone association. In the course of attending these classes, I befriended several young women engineers who helped me to negotiate access to their worksites in pattern-making rooms and product development departments at three different firms. From these new "value-added" sites in the industry, I was able to document how gendered meanings of skill and class position were drawn upon to restructure the labor process in the small handful of Dominican firms that were shifting from simple assembly to more complex production.

To study the livelihood circuits of retrenched garment workers, I followed Ferguson's (1999) longitudinal method in his study of displaced miners, adjusting my method and approach in the context of a weakly organized, mixed-gender factory workforce.[19] I focused my study on workers from a single firm that I call IA Manufacturing.[20] Two union federations, each with an affiliate union at the company, helped me to contact workers. Consequently, my key informants were former union activists who facilitated relationships with their former co-workers, both former union members and non-union workers. My interviews and repeat visits with former garment workers took place over the course of six months in two urban neighborhoods next to the trade zone and in three rural towns, or *campos*, where some workers had returned. In the initial phase, I conducted mostly individual and some small group interviews with workers about their work and migration histories, and their current strategies including sources of income, dependents and family obligations, amount of severance pay, indebtedness, and experiences in the labor market. My ability to keep track of workers' trajectories following these initial interviews rested on the deeper relationships I forged with key contacts. As evident in the opening vignette, Ambrosina and other key informants were also organic community leaders and problem solvers who maintained strong connections to their former co-workers, especially those from their hometowns, most of whom had found their trade zone jobs through these very ties to neighbors and kin in their *campos*.[21] Over a period of six months, I made multiple visits to their homes in Santiago and to their *campos*. Ambrosina and others helped to arrange follow-up interviews and would keep me updated on the whereabouts of their former co-workers.

The final site of this study is the border town of Ouanaminthe, Haiti, just across the river from the Dominican town of Dajabón. Intrigued by the claims of Dominican owners and US buying agents that Haiti was the future for their industry, I gained access to a new trade zone on the border situated between the two towns, owned and operated by the largest Dominican garment firm, which I will call Dominican Textile. I visited the zone on two occasions, spending several days based in the Human Resources department. Eventually, I moved to the town of Ouanaminthe in the summer of 2007 to better understand local dynamics. I conducted interviews in French, Haitian Creole, and Spanish, with the assistance of Mr Delien Blaise as a research collaborator and as an interpreter for the first two languages. It became clear that the conflict over the trade zone development was just one piece of a tumultuous local trajectory of change since the fall of the Duvalier dictatorship in 1986. What my industry informants in Santiago, supported by their US clients, saw as an obvious "greenfield" site for the global factory, community leaders in Ouanaminthe described as a dramatic urban expansion fueled by mass immigration and marked by the near complete collapse of infrastructure and services, setting the stage for the conflicts that would surround the new factories.

My research in Haiti took place before the devastating earthquake of 2010 that practically leveled the capital city of Port-au-Prince and the wider southern region. In the wake of the earthquake, the patterns of migration, trade zone development, and local tensions in the north that I explored in 2007 took on new dimensions. As international donors sought to demonstrate reconstruction through job creation and foreign direct investment, the pace of global factory creation accelerated and the northern region was constructed as a key growth area for transnational apparel companies. I consider these events in relation to my broader work on the garment industry's recent restructuring to show the power of the global factory as a development discourse not only to erase the legacies of capital investment in, and disinvestment from, Haiti, but also to foreclose the possibility of imagining alternative models of rebuilding the country. My focus on these three sociospatial locations contributed to what I now see as an important gap in my study: the state enters into my account through development plans and trade policies but its role in managing dis/investment – particularly in the case of the Dominican Republic – remains under-theorized.

My multi-sited approach was deeply influenced by work in anthropology, geography, and sociology over at least the past two decades to grapple with the question of where to locate "the global" in order to study "it" (cf. Tsing 2005).[22] As the idea of fieldwork became dissociated from the notion of the field as a bounded site "containing" a unique, hermetic culture, critical scholars encouraged a rethinking of the practice of fieldwork through feminist notions of location and situated knowledge (e.g., Gupta & Ferguson 1997). This displacement of the field as a cultural container has opened up the space for "global ethnographies" and "multi-sited fieldwork" made up of a series of engagements with different sociospatial locations linked not solely to different localities (i.e., sites), but more importantly, attentive to the multiple social relations that make up any particular place (Massey 1993). At the heart of such a formulation is the notion that all places are constituted by unequal power relations (Massey 1993), and that the researcher's own social location is not static, but rather is produced through interactions across hierarchies of difference, which are themselves always social and spatial.[23]

Because my research involved "mobile positioning" (Marcus 1998), I negotiated different relations of power depending on my engagements with subjects across intersecting hierarchies of race, class, gender, and national provenance. For example, Dominican academic colleagues were quick to point out that a Dominican graduate student would never have gained similar access to firms as I had, clearly based on my North American provenance and my racialization as white.[24] My North American-ness also raised suspicions. Managers and owners were well aware of, and highly sensitive to, criticisms of labor abuse in the garment industry. Moreover, the US labor movement had long supported Dominican union federations in the latter's efforts to organize the industry. My own previous involvement in the North American anti-sweatshop movement heightened my awareness of these tensions. On three occasions, managers mused about the status of my research and whether or not I might be infiltrating their businesses to expose them to the US consuming public.[25] In time, I was able to gain more unsupervised access through my relationship with the engineers from the Textile Center, but managers' and owners' suspicions about my intentions never fully dissipated.

Several predicaments of my mobile positioning arose – with the industry association, on the one hand, and with lower-level engineers and workers, on the other – that contributed significantly to my understanding of the politics of export restructuring *and* those of a multi-sited study. After completing the study of garment firms, facilitated graciously by the trade zone association's director, I left the association and began to spend time in working-class neighborhoods near the zone. Every few weeks, I dropped by the association to maintain my relationship with the director. On one of these visits, I mentioned that I was interviewing unemployed workers about their livelihood strategies after garment work. Following that conversation, the director began calling me to request my support in recruiting unemployed workers to speak about their experience to the local and national press. These requests signaled the employer association's dilemma: garment owners in the region had been highly effective in squashing strong independent union formation. Atomized labor relations meant that no significant, legitimate organizations existed to represent workers' interests during the industry's dramatic restructuring, a moment when, in other contexts, strong unions with the capacity to mobilize their base could have bolstered owners' efforts to gain government support for the sector. I was candid with the director about why I would not facilitate contact with the ex-workers who were participating in my research, since I would be using my mobility to assist in the sector's lobbying efforts. My relationship with the director was distant thereafter, but the exchange helped me to understand the precariousness of owners' efforts to make broader social claims in support of the industry.

This example shows how the mobility between research sites and social locations can make the relationship of unequal power inherent in "participant observation" of any kind more acute. If we disentangle participation from observation, and from the attendant contradictions that the phrase "participant observer" too easily neutralizes as an academic norm, "participant observation" can also be understood as a kind of problematic waiver held by the researcher to accrue social debts that are, from the outset, insoluble within the space-time of research encounters. As James Clifford argues, "participant observation is a kind of hermeneutic freedom to circle inside and outside social situations" (1997: 23). In a project that is multi-sited, the circulation between sites and social locations allows the researcher to defer these debts indefinitely. As in the case of the director's unspoken assumption that I should assist in the association's lobbying efforts in exchange for having facilitated my interviews with employers, other participants in my research contested my problematic waiver as well; in contrast to the director, these challenges came from their social positions of relative powerlessness. On many occasions, unemployed workers and engineers asked me to assist them in their (often transnational) strategies to find work, visas, and loans. These encounters demonstrated the incredible weakness of state-sponsored welfare to address Santiago's employment crisis, and thus the centrality of constructing social networks of support – and especially networks that could link to the global North – in order to navigate the precarity that accompanied mass unemployment. In response, I worked closely with organic community and union leaders who had much experience at finding fixes in the context of scarce resources and limited institutional capacity.[26]

In short, my efforts to forge a critical fieldwork practice were caught within persistent power differentials of knowledge production between North and South. This inequality is of course inseparable from the very relations of exploitation, exclusion, and livelihood spanning the global North and South that I explore in this book. Just as there can be no privileged outside to development discourses and progress narratives, so too can there be no escape from the relations of power that both structure the unequal production networks of the global economy *and* my efforts, as a researcher from the relatively more powerful consumer "end" of those networks, to destabilize them through my own displacement to sites of production and devaluation.

## Chapter Outline

In the chapters that follow, I bring together stories of worker livelihood strategies, union struggles, labor process restructuring, and policy and planning to unsettle the development discourse produced by the global factory in order to recuperate a politics of uneven development. I begin with an analysis of how the apparel industry became the centerpiece of trade policy and development narratives in the circum-Caribbean in the late twentieth century (Chapter 2). The chapter explores the gaps and failures of development accounts, which attributed the mushrooming of export factories either to the inexorable forces of globalization or to the entrepreneurial spirit of a particular provincial elite. My discussion leads into a feminist analysis of garment production restructuring in the Dominican Republic (Chapter 3). If the early garment boom relied upon low-wage feminized labor, I show how the subsequent reorganization of the industry mobilized gender to renew conditions of capitalist accumulation through the expulsion of feminized workers. I then turn from the factory floor to World Bank reports that interpreted these changes by revising the discourse of global production as development. This revised development narrative relied upon gender norms to rationalize the process of production restructuring – and labor devaluation – as a necessary stage in the country's progress. Chapter 4 reconsiders the global factory through the lens of precariously employed Dominicans who must navigate the boundary between wage and non-wage work. I argue that former trade zone workers' search for work in the wake of the garment industry's employment collapse is inextricably linked to their strategies to forge positions of social worth in and through rural and urban gendered and racialized spaces.

If export restructuring in the Dominican Republic functioned largely through the retrenchment of workers, in Chapter 5 I discuss the efforts of a Dominican company and its US corporate and state sponsors to transform the island's border area into a profitable margin for capital. The extreme uneven geographies between Haiti and the Dominican Republic created limits to this idealized "spatial fix" (Harvey 1999 [1982]), however; workers and managers cycled through these factories in search of opportunities that would allow them to reproduce their social positions and to meet their basic needs. In retrospect, border production represented a tenuous and small-scale attempt to reintroduce the global factory to Haiti. These efforts were radically scaled up in the wake of the country's devastating 2010 earthquake, the subject of Chapter 6. Key continuities existed with the export manufacturing strategy that swept the country in the 1970s, while the contemporary reconstruction discourse invoked an emboldened security imperative that cast the global factory as necessary to contain Haitians as a "black threat."

Several narratives of crisis run through these chapters and circulate widely in mainstream discourses of the Caribbean, from the sovereign debt crises of the early 1980s, which boosted the export manufacturing model, to the crisis narrative that justifies the global factory as a form of development in post-earthquake Haiti, to the relatively under-reported sovereign debt crises that have swept the Caribbean since the onset of the 2007–2008 financial meltdown sparked by the US housing market bubble. Chapter 7 reflects upon the pervasive trope of crisis in the Caribbean and argues that such dominant crisis events translate sociospatial change into sequential teleological narratives. Drawing on my fieldwork, Jamaica

Kincaid's poetic account of crisis, and contemporary reparations struggles in the Anglo-Caribbean, I offer three alternative framings of crisis that aim to unsettle Eurocentric temporalities of development and to cultivate solidarities around the political and ethical dimensions of uneven development. In the conclusion, I offer the reader some brief reflections on the implications of this text for our understanding of the role of labor in global production, variegated neoliberal regulation, and uneven development within and between countries of the global South.

## Notes

1  See Bair and Werner (2011) for an extended critique of this issue, which we have called the "inclusionary bias" of global production network studies. In contrast, we develop the notion of disarticulations as the study of constitutive inclusions and exclusions that make up any given production network (see also Bair et al. 2013).

2  The global North/South distinction I use here signals the historical and present-day divisions of labor and wealth both globally and in the Americas, where rent and surplus extracted from south of the Rio Grande concentrate in and through centers of accumulation and finance in the United States, Europe, and increasingly, Canada. Nonetheless, as I argue below, these categories *do not presume* a fixed geography. Rather, at the heart of my inquiry is an exploration of the relational and on-going construction of North–South divides as we witness the acceleration of uneven development and the fractured sociospatial divisions that are constructed through this process (see also Mohanty 2003; Sheppard & Nagar 2004).

3  I provide numerous examples of how the global factory as discourse operates throughout the text with respect to the Caribbean. For an empirical treatment of the industrialization-as-development discourse in terms of income and wealth inequality, see Arrighi et al. (2003). The authors demonstrate the divergence in national incomes worldwide, despite an apparent convergence in the proportion of manufacturing activity. Their article was the subject of debate with Alice Amsden (2003).

4  Indexed to constant 1968 dollars (Pinto Moreira 2010). GDP is the sequentialist development indicator *par excellence*. I use it here to demonstrate how the comparative case between Haiti and the Dominican Republic is made, which I subsequently unravel below and throughout the text.

5  I return to this argument at length in Chapter 6.

6  Thus, for Diamond, the comparison serves to defend his work against those who would consider him to be an environmental determinist by showing how "the environment" (understood here as first nature, or nature as an exogenous backdrop to social processes) shared by the two countries did not determine their divergent development trajectories. Culture, instead, as homogeneous and relatively stagnant, takes nature's place (Sheppard 2011; see Holmes 2010 for a refutation of Diamond based upon his research

on Dominican conservation policy). And while "hispanidad" generally functions as the debased cultural "other" of anglo/protestant capitalism, in the relativist logics of this comparison, the Dominican Republic serves as the more rational capitalist actor. Beyond this specific example, as I explore below and in Chapter 2, such cultural comparisons are also racialized, since hispanidad is constructed as a form of whiteness.

7   I discuss the significance of this inaugural debt in the conclusion.

8   For a comparative review, see Martin et al. (2002).

9   Trouillot used these terms to characterize Caribbean societies and the problem they posed for an anthropology that sought "traditional" or "native" societies. The Caribbean, Trouillot wrote, "has long been multiracial, multilingual, stratified, and some would say, multicultural" (1992: 21). On the claim of modernity, Mintz provocatively argued that the Caribbean was the site of the first modern, industrial societies. By the seventeenth century, the region's plantations were forward integrated enterprises that merged field and factory and coordinated a complex labor process that combined skilled, semiskilled and unskilled labor, free and slave, all regulated by a strict time clock (Mintz 1985; see also Williams 1994 [1944]).

10  In the English language, the radical work of André Gunder Frank (1967), and the more reformist approach of Cardoso and Faletto (1979 [1971]), are best known within this tradition. Numerous other scholars, however, such as Anibal Quijano, José Nun, Theotonio Dos Santos, Ruy Mauro Marini, and Celso Furtado carry as much or more weight in Latin America. Gunder Frank would later become closely associated with world-systems theory (see below) and the work of Immanuel Wallerstein following his exile from Chile after the 1973 coup against Allende. For a review of Latin American dependency thinking in English, see Kay (1989); for a discussion of the distortion of dependency thinking in English-language scholarship, see Slater (2004); on the distinctions between the Marxist dependency tradition and world-systems theory, see Sotelo Valencia (2005).

11  Specifically, world-systems theorists coined the term "commodity chain" as a "network of labor and production processes whose end result is a finished commodity" to describe the emergence and expansion of a worldwide division of labor dating from the end of the fifteenth century (Hopkins & Wallerstein 1986: 159; see Bair 2005).

12  In geography, see Dicken et al. (2001), Henderson et al. (2002), and Coe et al. (2004).

13  In the literature, the power exercised by lead firms over the chain is called firm governance (Gereffi 1994; see Gereffi et al. 2005 for a revised version of the concept). The possibility for firms to shift to higher value nodes is called upgrading (Gereffi 1999; Humphrey & Schmitz 2002). A firm may upgrade by, for example, increasing the range of activities that it performs for a client, or it may upgrade by increasing efficiency/lowering production costs in order to ensure continued orders from a lead firm that is benchmarking it against other suppliers. These core concepts are linked because the governance structure of a chain shapes the upgrading prospects of suppliers – a process that is by no means a guaranteed, or even likely,

outcome of participation in global production networks (Kaplinsky 2000). I am grateful to Jennifer Bair for her guidance through the intricacies of this literature.

14    Arrighi and Drangel, for example, explicitly sought to reinterpret uneven development through the workings of production networks, a theoretical project central to my own approach. The contemporary literature, from Gereffi onwards, has substituted the world-systems focus on uneven development for an inquiry into firm strategies to upgrade, conflating firm upgrading itself with development (see Werner et al. 2014). Geographers have contributed a partial corrective to firm-centered analysis by emphasizing regional development (e.g., Coe et al. 2004). This scholarship has largely focused on regional incorporation into production networks, however, thus largely ignoring the dynamic reproduction of uneven development in the contemporary global economy.

15    Federici more effectively demonstrates the centrality of gender to this process, a category that is tangential in Quijano's formulation. While Quijano's notion that coloniality shapes the distribution of the wage – centered upon Europe – is useful, the centrality of the wage and a singular notion of Europe is a significant weakness in his account. I am grateful to Massimiliano Tomba for pushing me on this point.

16    For an in-depth discussion of Quijano that situates his work within Latin American dependency/world-systems and postmodernism debates, and critically examines how Quijano's ideas have circulated in the Anglo-American academy, see Restrepo and Rojas (2010). One of their critiques of Quijano is his reduction of the notion of power to one of domination, marginalizing other notions of power that examine subject formation (e.g., hegemony and the role of consent in Gramscian-inspired formulations). In English, one example of a revised world-systems approach that troubles the notion of domination in post-emancipation societies is Michaeline Crichlow's *Globalization and the Post-Creole Imagination: Notes on Fleeing the Plantation* (2009). Below, and in Chapters 4 and 5, I disrupt the notion of coloniality as exclusively a relationship of domination through engagements with how current and former garment workers navigate coloniality and reshape its contours in particular ways.

17    The *encomienda* system granted Spanish colonists the right to extract labor and tribute from indigenous peoples, effectively producing populations that were "rightless and landless" (see Wynter 2003). For a powerful account of the shifting relations of production of banana workers in the twentieth century from full-time wage employees on banana plantations owned and controlled by US multinationals, through land reform and a period of production as independent producers, to precarious day laborers on contract farms selling to the modern-day incarnation of those same multinationals, see Striffler (2002).

18    Hall is worth quoting in full on this point: "One needs to know how different racial and ethnic groups were inserted historically, and the relations which have tended to erode and transform, or to preserve these distinctions

through time – not simply as residues and traces of previous modes, but as active structuring principles of the present organization of society" (1980: 339). The specificity of formations of social difference is crucial in any study that addresses the particular articulation of racism and exploitation at work in Haiti and the Dominican Republic. Again, to follow Hall, we conflate racialized forms of labor devaluation in the Americas as "essentially the same" at great peril, especially if we read these forms as mere variations of US racism (see also Trouillot 1995).

19  In 2003, before the downturn, 9 out of 531 trade zone firms had collective agreements. At the time of my research, no collective agreement had ever been signed with a trade zone enterprise in the Cibao region. The unions that existed in the Santiago trade zone represented small groups of workers who often hid their affiliation for fear of management reprisal. Moreover, Dominican labor law permits the formation of multiple unions in a single enterprise, further dividing the workforce.

20  Prior to closure, IA Manufacturing had at least four or, by one account, as many as nine registered unions.

21  Recruitment into export factories via rural ties is a very common occurrence in many countries. Through my fieldwork, I found very specific campo–factory links. In one case, for example, the factory ran a direct bus on the weekend to and from a relatively remote campo located four hours away in the province of San Francisco de Macorís by direct bus, a trip that would have taken considerably longer if workers had had to rely on public transportation.

22  My method is influenced by three kinds of scholarship: ethnographies of global connections in various disciplines, including Clifford (1997), Gregory (2007), and the work of Michael Burawoy (1998; Burawoy et al. 2000) and his elaboration of the extended case study method; ethnographies of production and regional formation, especially Chari (2004), Hart (2002), and Narotzky and Smith (2006); and feminist critics of the globalization literature who insist upon grounded studies of global industries that decenter masculinist conceptions of labor such as Freeman (2000), Silvey (2000), Wright (2006), Cravey (1998), Salzinger (2003), Ramamurthy (2011), and Collins (2003), as well as classic works in this field including Elson and Pearson (1981), Fernández-Kelly (1983), Ong (1987), and Wolf (1992).

23  See Rich (2001 [1985]), Haraway (1991), Katz (1994), and Mohanty (2003); see also Freeman and Murdock (2001) for a review of this literature in studies of Latin America and the Caribbean.

24  For example, in several interviews, owners and managers clearly presumed a kind of "white solidarity" in their candid, negative comments about East Asian garment investors, exacerbated by the emergence of China's export capacity. If my racialization and provenance afforded me some limited access to garment managers and owners, my gender and youth often circumscribed the depth and quality of information shared by these largely male business elites.

25  Accusations of spying or covert sympathies either for labor or for the competition are not an uncommon part of fieldwork in export firms. See, for example, Carla Freeman's (2000) introduction and Wright (2001).

26  Recall, for example, Ambrosina's pledge to help her neighbor in Los Almácigos, Leidy, to get social security payments for her child who had been born while she was still entitled to this benefit through her job at IA Manufacturing.

# 2

# Two Stories of Caribbean Development

## Garments-as-Globalization and Garments-as-Regional Entrepreneurialism

## Introduction

This chapter considers how the apparel industry was linked to notions of development in the circum-Caribbean and in the Dominican Republic over the course of the last 30 years of the twentieth century. I present the reader with two accounts of development: garments-as-globalization and garments-as-regional entrepreneurialism. The first story recounts how US trade policy, driven by late Cold War geopolitics and shaped by flexible production strategies of multinational corporations, vigorously promoted the industry in the circum-Caribbean. In the wake of the limits of national policies of import substitution and subsequent debt crises, and the repression of radical projects for socialist transformation, garment assembly became a potent symbol of globalization for both the promoters of neoliberal reforms and their critics. Some scholars and numerous policy proponents seized upon the mushrooming of factories and skyrocketing manufacturing export values to celebrate the globalization of production, while others took the same phenomenon as evidence of a pernicious race to the bottom. In general, these accounts considered the rise of a macro-regional garment production network as the outcome of forces stemming from the global North (both state and capital), and failed to theorize the significance of place-specific contours of disinvestment and restructuring.

If this first story of development framed the growth of the garment industry in the circum-Caribbean as either a felicitous or a pernicious turn

*Global Displacements: The Making of Uneven Development in the Caribbean*, First Edition. Marion Werner.

towards the seemingly inexorable forces of globalization, the second story of development that I recount here interrogates the industry's particular trajectory in the Dominican Republic. The consolidation and deepening of garment production in the country's northern Cibao region stands in sharp contrast to several other Caribbean states — and the rest of the Dominican Republic – where the industry failed to catalyze significant backward and forward linkages and moved on quickly to Central America and Mexico. In this account, the garment industry's consolidation in the Cibao serves as evidence of a particular kind of regional entrepreneurialism as the inheritor of a developmental agro-export tradition rooted in the northern region's "tobacco culture." This regional culture stands in contrast to the weak presence of local capital and the predominance of foreign interests in the country's former sugar plantation zones in the southeast, the other site of the garment boom in the country in the 1980s.

Taken together, these two stories present themselves initially as distinct, opposed accounts. The garments-as-globalization narrative – whether told from the perspective of neoliberal promoters or detractors – ascribes explanatory power to the role of multinational capital and the US government. The spread of the garment industry is viewed as the result of slotting countries into externally defined positions in the global economy. The garments-as-regional entrepreneurialism account, on the other hand, attributes the relative success of the industry in the Dominican Republic to the role of a local fraction of capital rooted in a particular capitalist agrarian regional history. Both stories are as important as they are flawed. The first presumes to erase the particularities of uneven development and suffers from what Gillian Hart (2002) has called the "impact model" of globalization, or the presumption that places passively receive global forces and are shaped unilaterally by them. Both accounts ground their explanatory power in the actions of capital, either multinational or provincial, assuming that businesses respond unilaterally to changing market incentives and re-allocate land, labor, and capital accordingly. As I argue throughout this book, however, uneven development must be made. As a process that reflects, animates, and escapes the restructuring of value hierarchies in the global economy, uneven development is certainly not made by capital alone.

In the following section, I reconstruct the garments-as-globalization story. I describe for the reader the trajectory of garment industry growth in the circum-Caribbean, accelerated under the policies of US President Reagan in the context of the Cold War and the Latin American debt crisis. My aim is to signal not only investment expansion leading to the spread of hundreds of global factories, but also pervasive trends of disinvestment. Rather than understand disinvestment as evidence of failed development, I argue that factory mobility and flight created new geographies of uneven development that fueled the on-going and dynamic production of the macro-region as available for this type of labor intensive, low-wage factory boom. In the section titled "Dominican Counterpoint," I focus in on the specific trajectory of the export garment industry in the Dominican Republic, or the story of garments-as-regional entrepreneurialism. Among its circum-Caribbean neighbors, the Dominican Republic was the only early participant in the garment export boom that persisted well into the model's phase of consolidation and restructuring. I explain this relative stability, and the profound impact that the sector had on the country, through a brief account of the industry's rise and development in the northern Cibao region, where a relatively large number of local investors became owners of garment firms and created local linkages. The persistence of the Dominican garment export industry must be understood in part through the lens of these particular regional politics – stemming from long-standing regional differences in agro-export production – which ultimately shaped the trajectory of capital accumulation created through these new manufacturing exports.

In the conclusion, I return to the limitations of both these stories, which circulate as predominant narratives of late twentieth-century development in the circum-Caribbean. Both are relevant, but neither is adequate. On the one hand, the garments-as-globalization story both obfuscates uneven development and fails to account for the particularity of place. Relationships of investment and disinvestment are overshadowed by a focus on sequential development and a debate over successful or failed entrants into global production arrangements. The story of regional entrepreneurialism, on the other hand, offers a more nuanced understanding of the place-specific contours of global production restructuring, but the dynamics of uneven development remain opaque. The focus on the actions of local capital ignores the way that places are shaped through class relations and articulated with gender and racial difference. Moreover, accounts of regional entrepreneurialism tend to minimize the constitutive nature of restructuring and disinvestment. In short, while important for our understanding, neither story adequately accounts for the contested processes that make uneven development. This limitation motivates the chapters that follow.

## "Taiwanizing" the Caribbean? Garments-as-Globalization

The post-1970s trajectory of industrial promotion strategies in the circum-Caribbean can be traced to Puerto Rico's Operation Bootstrap and its dissemination by the plan's intellectual authors and implementers.[1]

Investment incentives adopted in 1947 prompted the growth of low-wage manufacturing in Puerto Rico. As a result, the garment industry grew rapidly, comprising between one-quarter and one-third of manufacturing jobs on the island from the mid-1950s to the mid-1970s, largely employing women workers (Safa 1995b). St Lucian-born development economist W. Arthur Lewis advocated the Puerto Rican model for the Anglo-Caribbean. Lewis (1949) argued that in order for insular Caribbean territories to overcome their dependence upon primary commodity exports, national elites would have to lure export-oriented multinational companies to their shores by offering generous fiscal conditions. In the 1950s and 1960s, in line with these ideas, several countries passed industrial promotion laws to establish industrial or trade zones with the aim of creating platforms for offshore manufacturing for export to the United States and Europe. The first of these efforts dates from Jamaica's 1956 Export Industry Encouragement Act, followed by Haiti in the early 1960s, and the Dominican Republic in 1968 (Heron 2004; 2006). Through these regulations, states designated special areas or an enterprise status that would attract export-oriented investment through benefits and exemptions favorable to foreign investors. But if such policies sowed the seeds for the later garment boom in the macro-region, during these early decades they failed to catalyze employment growth and inward foreign direct investment of this kind. The ideas of Lewis and the policies developed in mid-century Puerto Rico – focused as they were upon promoting industrial development oriented towards markets in the global North – ran counter to the predominant trend in Latin America and the Caribbean. For at least another decade, state priorities would lie with the promotion of import substitution industries geared towards domestic markets and production, a regulatory orientation that was inspired by variants of structuralism and dependency thinking and emboldened by the Cuban revolution.

In the 1970s, the combination of revolutionary movements, dependency critiques in the English Caribbean, and the detrimental effects of the oil shocks on Caribbean countries led to the election of socialist governments in Jamaica and Guyana, and the seizing of power by a Marxist-Leninist party in Grenada. Leftist movements were also mounting radical challenges to the predominant structures of dependent capitalism in Central America. In Nicaragua, the leftist Sandinista coalition successfully overthrew the US client-dictator Anastasio Somoza in 1979. In short, US hegemony in the region faced unprecedented contestation, a situation that would convert the region into one of the most violent fronts of the late Cold War over the following decade.

The 1970s yielded different fortunes for Haiti and the Dominican Republic, however, and this divergence was crucial to both countries' early positioning as garment exporters. In Haiti, power was transferred under the repressive dictatorship of Francois Duvalier to his son, Jean-Claude, or "Baby Doc," upon his father's death in 1971. In contrast to his father, Jean-Claude Duvalier embraced the export-oriented model more fully, and by 1972, 58 export assembly firms were operating in the Haitian capital, a number that grew threefold over the decade (DeWind & Kinley 1988; see also Trouillot 1990). Through assembly production of imported parts in the garment sector and other "light" industry including baseballs, toys, and electronics (DeWind & Kinley 1988), Haiti outpaced other countries in the region in its proportion of manufacturing exports, which reached 58 percent of the country's total exports by 1980 (Bulmer-Thomas 2003).[2] In the Dominican Republic, Joaquín Balaguer, the close ally and successor of Dominican dictator Rafael Leónidas Trujillo, was installed as president with the aid of a 16-month US military occupation. Surprisingly, Balaguer did not capitulate to the neoliberal reforms advanced by his US patrons, opting instead to create clientilistic ties through domestic industrial subsidies and protections, thus extending the country's import substitution orientation (see below). Nevertheless, a mix of private and publicly sponsored projects led to the creation of three trade zones by the end of the decade. The output from these new zones was significantly dwarfed by the export factories of neighboring Haiti.

When President Reagan came to power in 1981, he faced two seemingly unrelated challenges. On the one hand, doubling down on its commitment to the Cold War, the Reagan Administration pledged to restore US hegemony in the circum-Caribbean, a region that the US had considered "its backyard" since the days of Theodore Roosevelt.[3] On the other hand, the Administration faced growing calls for protection from US manufacturers as the country registered its most significant manufacturing trade deficit since 1917. Apparel and textile manufacturers, in particular, were boosting their campaign to reverse trade benefits granted to South Korea, Taiwan, and Hong Kong during the period of post-World War II reconstruction (Rosen 2002). The three countries represented a majority of textile and apparel imports at the time (Rosen 2002). Over the course of the decade, these two challenges would be twinned: the Reagan Administration would appease textile industrial protectionists and renew US hegemony in the circum-Caribbean in part through the subordinate incorporation of offshore firms and workers in garment piece-assembly for the US market. The result would be the creation of a macro-regional garment and textile production network, linking the United States, Mexico, and the circum-Caribbean, together with East Asian, US, Mexican, and circum-Caribbean capital, and hundreds of thousands of Mexican and circum-Caribbean workers.

Initially, the Reagan Administration sought to address primarily the first challenge – that of faltering US hegemony – by boosting military spending and economic integration with the region. In order to do so, Reagan recuperated the geographic category of the "Caribbean Basin" from the archives of Manifest Destiny and enacted the country's most wide-sweeping Caribbean-related legislation of the post-World War II period.[4] The Caribbean Basin Initiative (CBI), as the legislative package was called, encompassed a diverse set of trade, development, and military policies to serve US geopolitical and economic goals. The Reagan Administration argued for the passage of the legislation in order to strengthen ties between US and Caribbean capital, to promote private sector initiatives, and thus to keep the threat of Communism at bay. Nearly one-third of the 350 million dollars in funds appropriated by Congress under the CBI was destined for military interventions in El Salvador; the remainder was earmarked for "economic revitalization" in the rest of the region (Heron 2004). The trade component of the CBI, known as the Caribbean Basin Economic Recovery Act (CBERA), was directed towards "modernizing" Caribbean economies by promoting so-called non-traditional exports such as off-season fruits and vegetables (Heron 2004). For the insular Caribbean, this focus of economic development policy was meant to aid in the diversification of sources of foreign exchange away from sugar.[5] At the time, Caribbean nations were reeling from declining sugar prices and the contraction of US sugar import quotas.[6] Thus, in order to boost other exports, the CBERA granted one-way,[7] duty-free access to non-competing products from all insular Caribbean and Central American countries, with the exception of Cuba. Significantly, because the legislation framed garment and textile products as competing goods, these two sectors were excluded from duty-free importation and US development funds for their promotion.[8]

Despite their exclusion from the CBERA, apparel exports continued to grow from the region because US garment manufacturers were expanding their assembly operations there. This expansion was facilitated by two key pieces of trade regulation. The first was the Multifibre Arrangement, or the MFA. The emergence of export textile and garment producers in East Asia in the 1960s had created a consensus among big importers in the global North to introduce quota restrictions on textile and eventually garment imports as a special exception within the multilateral trade system, called the General Agreement on Tariffs and Trade, the GATT (Underhill 1998). Although nominally a temporary arrangement when signed in 1974, the MFA lasted for 20 years. Under the MFA, large importers – principally the United States, Europe, Canada, and Australia – had the right to establish quantitative restrictions on the imports of garment and textile products from any given producer country. As quotas were placed upon high export countries, companies relocated production capacity to countries that were not yet constrained by quota limits (Underhill 1998). As a result, by the early 1980s producers from more than 100 countries were exporting to the US market, making garment and textile by far the most "global" of manufacturing industries (Rosen 2002), albeit through well-defined networks controlled by particular firms.

The second, related piece of regulation that contributed to the growth of apparel exports from the circum-Caribbean was the easing of restrictions on the "re-importation" of goods assembled from parts made in the importer country, also called "outward processing trade," or OPT. In the US context, manufacturers were taking advantage of a special provision in the tariff code, known by its subheading 807 (later called 9802), that required duty to be paid only on the value added – mostly the cost of labor – on goods that were assembled from US components (Rosen 2002). Thus, although the initial CBI package excluded garments and textiles from duty-free import to the United States, under the 807 provision, garment importers were already integrating circum-Caribbean countries through the creation of assembly platforms. The resulting increase in garment exports (or "re-imports") was outpacing any of the other product categories formally promoted under the legislation. In effect, trade regulation was being transformed by the changing structure of the industry itself. In line with the broad shift towards flexible arrangements in global production networks that I described in Chapter 1, garment manufacturers used these trade provisions initially to locate directly owned factories in lower wage countries. Eventually, these firms disinvested from production altogether through outsourcing to Caribbean suppliers as well as to East Asian contractors operating in the Caribbean. These primarily South Korean, Hong Kong, and Taiwanese firms set up operations in the region in order to avoid MFA quota ceilings and to benefit from the 807/9802 value-added tariff arrangement.[9]

By 1986, the Reagan Administration recognized the potential to satisfy US textile protectionists and to serve the country's geopolitical interests by promoting apparel assembly exports from the Caribbean region. To achieve this goal, the US government implemented an enhanced "production sharing" model in 1987, called the Special Access Program, that guaranteed unlimited quotas, called Guaranteed Access Levels, and duty payments on value added only for garments made of US woven or knit cloth sewn in countries that were party to the CBI (Rosen 2002; Heron 2004).[10] This arrangement thus protected US textile producers facing the erosion of their sales from the importation of garments made from cheaper cloth produced in East Asia, while promoting low-cost, labor-intensive garment assembly using US-formed cloth in circum-Caribbean countries.

By encouraging the expansion of the garment industry under the new Special Access Program, promoters of the legislation argued that the provisions would help reproduce the "Asian Miracle" in the Caribbean, while also strengthening US-aligned governments in the region (Rosen 2002). Indeed, politicians and boosters argued that the garment trade and associated agreements would "Taiwanize" these exporting countries, drawing upon a commonsense understanding of garments and textiles as a first stage of industrialization repeated by territorial units.[11] For much of the late 1980s and 1990s, however, many academics and policy observers argued otherwise, predicting that these new exports would fall short of such grand expectations (e.g., Kaplinsky 1993).[12] Most obviously, the trade provisions actively discouraged increasing the local content of garments and thus the sorts of backward linkages, especially to textile production, that are associated with consolidating industrial production, since tariffs were levied precisely on the local content until the mid-1990s. Moreover, the arrangement created incentives for subcontract firms that were entirely dependent upon US capital for technology and inputs, while narrowly specializing in providing low-cost labor, in contrast to East Asian producers that learned from the start how to manage the entire process. While much can be learned from the debate over the developmental consequences of the region's new export orientation, its very framing rested on the notion that countries follow a sequence of development stages; the trajectories of most circum-Caribbean countries ultimately represented their failure to do so.

This particular debate about the role of garment assembly and other new exports in the region's development success or failure was inseparable from the broader debate over the turn towards neoliberal policy marked by the implementation of structural adjustment policies throughout Latin America and the Caribbean during the 1980s.[13] Faced with mounting debt obligations as a result of high oil prices, declining primary commodity prices, and an increase in interest rates, most countries in the region could not service their sovereign debts. As a condition of debt renegotiation, circum-Caribbean countries acceded to structural adjustment policies designed and enforced by the International Monetary Fund (IMF) that radically restructured their regulatory regimes in line with the priorities of multinational capital. This set of policy prescriptions came to be known as the Washington Consensus (Fine 2001; Rodrik 2006; see Chapter 1). The Washington Consensus measures that most directly supported the garment boom included the strengthening of rights and protections for foreign investors, the devaluation of domestic currencies, and the loosening of protective labor legislation. Other common prescriptions included the privatization of public companies, a shrinking of the public sector, and the liberalization of debtor country tariffs, eventually leading to the replacement of provisional so-called preferential arrangements like the CBI with fully bilateral trade agreements.[14]

The result of these extensive trade and macroeconomic regulatory shifts on the macro-regional garment and textile production network was nothing short of dramatic. By the end of the decade, hundreds of thousands of circum-Caribbean workers were incorporated within the industry's highly volatile and uneven production structure in part through the wholesale transfer of productive capacity from East Asia. Whereas in 1987, China, Hong Kong, Taiwan, and South Korea represented 56 percent of the import share in the US garment market, relative to 8 percent from CBI countries and 2 percent from Mexico, by 1998, these same Asian countries' share had dropped to 22 percent, while the import share of CBI countries and Mexico had risen to 24 percent and 15 percent respectively (Heron 2004: 93). In fact, by the late 1990s, so much production had moved to the circum-Caribbean and Mexico that large Asian producers were not filling their country quotas under the MFA at all (Underhill 1998). The largest garment exporters to the United States in the 1980s were Haiti, Jamaica, the Dominican Republic, and Costa Rica, together responsible for more than 80 percent of assembly exports from the CBI region (Heron 2004; see Table 2.1). The spurious assertions by US policymakers that US policy would contribute to making the Caribbean in the image of Taiwan ignored how the region

**Table 2.1**   United States apparel imports by macro-region, 1987 and 1998

| Source | 1987 (millions of square meters) | Import share (%) | 1998 (millions of square meters) | Import share (%) | Change 1987–1998 (%) |
|---|---|---|---|---|---|
| NICs, China | 3,040 | 56 | 2,853 | 22 | –6 |
| ASEAN | 709 | 13 | 1,469 | 11 | 107 |
| South Asia | 547 | 10 | 1,858 | 14 | 240 |
| CBI | 436 | 8 | 3,066 | 24 | 603 |
| Mexico | 134 | 2 | 1,985 | 15 | 1381 |
| All other | 593 | 11 | 1,655 | 13 | 179 |
| Total apparel imports | 5,459 | 100 | 12,886 | 100 | 136 |

Source: American Apparel Manufacturers Association (1999) in Heron (2004: 93). Reproduced by permission of Ashgate Publishing Ltd.

Note: Newly industrializing countries (NICs) refers to South Korea, Hong Kong, Singapore, and Taiwan.

would indeed become linked to East Asia: not reproduced in its image but as a subordinate recipient of Taiwanese, South Korean, and Hong Kongese capital.[15]

From an "impact model" perspective, garment export growth in the Dominican Republic was paradigmatic of these macro-regional trends. After protracted resistance to IMF structural reforms, in 1983 the Administration of Jorge Blanco acceded to IMF monetarist demands in the face of large fiscal and current account deficits. In order to receive a three-year, 400 million dollar loan, the Blanco government agreed to reduce public spending, cut food subsidies, restrain the money supply, implement a sales tax, and, crucially, liberalize the Dominican peso (McAfee 1991; Espinal 1995). Devaluation and austerity measures began in 1984. The resulting price spikes ignited the accumulated frustration of urban popular sectors already reeling from high inflation that had severely eroded wages over the previous two years.[16] A three-day nation-wide popular uprising – called *la poblada* – ensued, by the end of which at least 60 people had been killed and hundreds more injured.[17]

Both the Dominican government and the business community used *la poblada* strategically to advocate for market access and more aid from the United States by raising the specter of political instability in a country that President Reagan had described just one month earlier as "a beacon of freedom-loving people everywhere" during a visit by Dominican President Blanco to Washington to request more aid.[18] Over the next three years, macroeconomic, trade, and aid measures would coalesce further to spark a veritable boom in garment exports. In 1985, the government fully liberalized the peso and as a result, Dominican manu-facturing wages relative to average US manufacturing wages fell from 12 percent to 5 percent (Mortimore et al. 1995). The three trade zones that had been established in the late 1960s and early 1970s under the first export incentive law (i.e., Law 299 of 1968), which employed approxi-mately 20,000 workers before structural adjustment, quickly filled to capacity (Abreu & Cocco 1989: 142). Indeed, trade zone employment quadrupled between 1985 and 1988 (Abreu & Cocco 1989: 142), spurred by the new provisions of the Special Access Program. Dozens more trade zones were established and by 1990 official employment figures registered 130,000 workers (CNZFE 1999: 21), a more than six-fold increase in six years. Three-quarters of these jobs were in the garment assembly sector. The resulting impact on export figures was formidable: trade zone exports increased by 240 percent, surpassing the export share of agricultural commodities and metals – that is, "traditional" exports – which declined by 90 percent (CNZFE 1999: 21).

As more and more countries in the circum-Caribbean acceded to structural adjustment policies and sought to attract garment and other export manufacturing investment in order to generate sources of foreign exchange, competition within the region intensified. The policies that generated investment and new exports for a single country eroded the basis of these shifts when adopted by multiple producers.[19] Competitive currency devaluations are the most oft-cited example of this pernicious dynamic. As a given state devalued its currency, neighboring states competing for the same investment faced extraordinary pressure to do the same, pushing wage rates down across the region and eroding the purchasing power of workers. Over the decade, the effective devaluation of circum-Caribbean currencies was more than 20 percent relative to their exchange rate with the US dollar in 1980 (Kaplinsky 1993; also Mortimore & Peres 1998: 52).[20] The cost of not following the trend of competitive devaluation was disinvestment. In Barbados, for example, a government commitment to a stable currency "priced" Barbadian labor out of this type of investment, leading to the swift exit of garment and other assembly jobs by the end of the decade (Freeman 2000).

Such competitive pressures only increased in the 1990s. By the middle of the decade, El Salvador, Honduras, and Guatemala became important sites of garment production. More significantly, Mexico became a major garment exporter with the passage of NAFTA, which provided greater market access for Mexican producers to the US market in two ways. First, NAFTA eliminated tariffs and quotas on garments produced in the region.[21] Second, to qualify for this duty-free treatment, garment exports could be produced with yarn and fabric from any signatory of the agreement, allowing for the incorporation of Mexican textiles in garment exports to the United States (Heron 2004; Bair & Dussel Peters 2006).[22] In addition to market access, intra-regional competition continued to stiffen on the basis of currency value. While circum-Caribbean currencies steadily appreciated with respect to the US dollar during the 1990s, the devaluation of the Mexican peso in December 1994 and January 1995 effectively cut hourly wages in the Mexican *maquila*, or export-processing, sector by fifty percent (Heron 2006: 270). Thus, while the initial passage of regional trade provisions and structural adjustment policies in the circum-Caribbean, combined with quota restrictions on East Asian producers, facilitated a massive shift in production capacity from Asia to the macro-region, the subsequent intra-regional competitive dynamics led to the enrolment of new investment sites that largely grew at the expense of factory flight from neighboring countries.

The final key regulatory change to impact the global organization of the industry was the passage of the Agreement on Textile and Clothing as part of the Uruguay round of multilateral trade negotiations in 1994. Signatories to the newly formed World Trade Organization (WTO), which succeeded the GATT in 1995, agreed to a phase-out of quotas regulated under the MFA over a 10-year period, from 1995 to 2004. The phase-out was seen as a concession to countries in the global South and Eastern Europe, won in part through their acceptance of new trade disciplines like intellectual property and investment measures as part of the multilateral trade agenda (Underhill 1998, Chang 2007). From the perspective of small producer countries like those of the circum-Caribbean, the end of quotas promised a scaling up and further intensification of the competitive dynamics of the industry that already animated the investment and disinvestment patterns that I have described. Production in the macro-region had hitherto depended upon a "supply gap" created by constraining imports from lower cost producers in Asia, especially China, Bangladesh, Indonesia, and Vietnam by the late 1990s. With the end of the global quota regime beginning in 2005, circum-Caribbean and Mexican garment exporters would compete directly with their East and Southeast Asian counterparts. The result would be a large-scale shift in production capacity and jobs from the former to the latter. Indeed, in the initial adjustment to the post-quota garment trade (2004 to 2008), the US market share supplied by circum-Caribbean and Mexican producers fell from 22.8 percent to 13.3 percent, while the share supplied by large Asian producers grew from 26.1 percent to 50 percent (Lopez-Acevedo & Robertson 2012: 62–63).

The proliferation of garment assembly factories in the circum-Caribbean and Mexico that I have described here is widely attributed to a turn in the regulatory orientation of these producer countries associated with globalization in the late twentieth century. The architecture of trade agreements and structural adjustment policies, driven by late Cold War logics and the subsequent triumph of market orthodoxy, created the context for the macro-regionalization of the industry, while also placing these global factories at the center of debates on capitalist development and its failures. And yet such accounts are insufficient for understanding how uneven development is made since they presume places in the global South are simply available for capital investment and disinvestment without interrogating the social struggles and local histories that shape this process. Moreover, disinvestment is generally analyzed in terms of either regulatory failure – as in the case of competitive devaluation that created a pernicious race to the bottom in the region – or state or local capital failure – as in the failure to catalyze backward and forward linkages. Disinvestment and restructuring are rarely viewed through the lens of uneven development – that is, as processes that reconstitute the unevenness upon which production networks are premised.

In the next section, I turn to the specific case of the Dominican Republic. The country has received extensive scholarly attention since it was not only an early garment exporter in the 1980s, but the Dominican

Republic was also the only country among the early exporters that remained a significant producer well into the 1990s and the early 2000s. While the debate over the Dominican Republic's adoption of an export-oriented model reproduces the success-versus-failure framing that we have seen, the sector's particular regional geography introduces a hitherto marginal dimension: the role of local capital and regional trajectories of economic change. Such an account adds to our understanding of the dynamics of uneven development, even as it remains incomplete.

## Dominican Counterpoint: Garments-as-Regional Entrepreneurialism

Regional histories and, in particular, the agrarian roots of industrial change, are key dynamics through which we can understand the making of uneven development. As I argued in Chapter 1, the abstract fields of uneven development – that is, which firms do what where *and* how value hierarchies of labor articulate with coloniality – combine into concrete determinations of place production. Regions are conformed by layered histories, class dynamics, and struggles over culturally constructed meanings of value and identity that shape the outcomes of new rounds of investment, disinvestment, and re-regulation, just as they are reshaped by these forces. In short, the circum-Caribbean garment boom was not solely the product of decisions made in Washington and in corporate headquarters of multinational corporations that determined a new set of economic incentives and production practices. Rather, the location, duration, and structure of the shifting value hierarchies of production articulated with particular regional processes, eventually producing new geographies of uneven development as certain places became sites for the production of greater or lesser capitalist value, and other locales were provisionally constructed as outsides to these networks.

In the Dominican Republic, the new export model articulated with existing regional differences rooted in late nineteenth- and early twentieth-century agrarian relations. Over the course of this period, two distinct agro-export zones were established. US capital financed sugar plantations on the country's southeastern coastal plane, which expanded rapidly, growing six-fold between 1905 and 1925 (Turits 2003). The US occupation from 1916 to 1924 consolidated the sugar enclaves and foreign ownership and control over them. The parallel US occupation of Haiti – from 1915–1934 – established a pattern of migrant labor that would endure for much of the twentieth century. Haitians emigrated in high numbers from the more densely populated and settled western third of the island. They were propelled by land seizures for US capital interests, which combined with a protracted agrarian crisis in the country, increasing the numbers of landless and land-poor. Rural Haitians were recruited through direct campaigns and programs in order to work on US-owned sugar plantations in the Dominican Republic and Cuba (Castor 1971).

The fortunes of the northern Cibao valley were distinct from those of the southeast. In the late nineteenth century, traders and merchants – including many European and Levantine immigrants – began to commercialize primarily tobacco, as well as coffee and cacao, for export to the European market (Baud 1995). But cultivation of cash crops constituted a relatively minor use of the region's land, much of which remained an open frontier dominated by swidden agriculture, pastoral grazing, and collection (Turits 2003).[23] Access to land created relatively enduring patterns of semiproletarianization that favored the autonomy of rural dwellers who were not averse to participation in markets, but were not compelled to do so. In the Cibao, farmers cultivated tobacco on small plots while also producing for their own subsistence. Dominican peasants participated as seasonal laborers on sugar plantations in the early phase of mono-crop expansion, but quickly removed their labor when prices for cane cutting dropped following the late nineteenth-century sugar crisis (Turits 2003). Eventually, the Haitian migrant labor regime filled this unmet labor demand on sugar plantations and continued to do so for much of the twentieth century.

The distinction between agro-exports produced through foreign-controlled *latifundios*, or plantations, versus *criollo* elites commercializing goods produced by independent farmers on *minifundios*, or smallholdings, was the object of political and cultural analysis in the country long before Fernando Ortiz's famed comparison of the cultures of sugar and tobacco in his book *Cuban Counterpoint* (1995 [1947]). Citing the ravages of the emergent sugar enclaves in the country's southeast, a growing number of liberal elites, many from the Cibao, were already conceiving of and advocating for an alternative modernity for the Dominican Republic in the late nineteenth century.[24] These liberal reformers held up the male surplus-producing peasant as the ideal citizen of a modern nation-state, while denouncing the moral and political hazards of seasonal work on plantations and growing joblessness, hunger, and moral decrepitude in urban centers (Turits 2003). Skeptics of this liberal agrarian vision evoked a dual, racialized construction of rural subjects: the market-integrated peasant, idealized as "more European," was but a small and relatively insignificant presence in contrast to a "black" rural majority organized into autonomous, often mobile communities and settlements, descendants of slaves, who were unruly and "unproductive" (González 1993; González de Peña 2004).[25] The alternative regional modernity of the liberal reformers was inseparable from a project of identity that sought to construct the Cibao as the locus of *dominicanidad* – or Dominican-ness – an identity that projected a unity of interests between a domestic, commercial, developmentalist elite and a productive peasantry, which together could mitigate the pernicious influence of foreign capital and construct the Dominican nation (González 1993; Yunen 1985).[26] While this discourse tended to essentialize tobacco cultivation as inherently a harbinger of more stable and democratic social structures, Hoetink (1980) argues that in fact the determining factor was not the crop but rather the historical context within which tobacco exports were established. Tobacco cultivation expanded during a period when neither landed property nor urban commercial functions were monopolized, a condition that precipitated a greater distribution of land and commercial profits (Hoetink 1980). The result was indeed a social structure that differed significantly from the rigid hierarchies of plantation enclaves in the southeast. "In the Cibao countryside," historian Michiel Baud explains, "the classes lived side by side and were in regular, sometimes intensive contact with each other. Elite children often entertained friendship with age-peers within the peasantry, which could be the base for a life-long mutual esteem" (1997: 365). Such rural class proximities likely also tended to form complex relations of paternalism (see, e.g., Narotzky & Smith 2006). In the twentieth century, the peasant producer ideal would be linked more directly to a racial ideology of "hispanidad" and whiteness as the Dominican state embarked upon its modern project of nation-making.

For the first two decades of his 31-year rule, Rafael Leonidas Trujillo drew at least in part upon the elite liberal vision of an alternative agrarian modernity (Turits 2003). The absence of an established planter class, stemming from the country's marginal position relative to colonial Saint-Domingue followed by the widespread abandonment of the island by elites after the Haitian revolution, permitted the dictator to consolidate a political base through mass property-making via land distribution (Hoetink 1982; Moya Pons 1986; Turits 2003). The result was the most extensive land reform in Latin America and the Caribbean at the time: one-third of male rural producers became small property owners of average-sized plots of 2 hectares on 10 percent of the country's occupied land (Turits 2003).

In the context of the rise of fascism in Europe, Trujillo combined the alternative smallholder agrarianism advocated by the liberal elite with a robust doctrine of anti-Haitianism and negrophobia. The dictator constructed Haitian peasants and peasants of Haitian descent as internal enemies to be eradicated from the countryside, precipitating the massacre of an estimated 15–20,000 people that state forces identified as Haitian, primarily in the border region and the wider Cibao valley in 1937 (Derby 1994; Turits 2002).[27] Additionally, through tax obligations combined with infrastructure development and subsidized credit, the regime's productivist ideology tied citizenship to settled agricultural cultivation and was extremely punitive towards those who did not conform to these dictates (San Miguel 1999; Turits 2003). The real subsumption of the peasantry to the market was recast in racialized terms. Key anti-Haitian ideologues of the regime, especially Manuel Peña-Batlle and Joaquin Balaguer, erected the myth of the white *campesino* of Spanish origin against the rural vagrant corrupted by racial mixing and Haitian cultural influence (San Miguel 1997). Balaguer (1984) would give this dualism a spatial form: the locus of a fabled "pure," lighter-skinned *campesino* who embodied Dominican culture was the mountain range of the central Cibao, while those peasants "corrupted" through integration with Haitian laborers resided on the border and in the southeast sugar region.

In the 1950s, Trujillo pursued a policy of import substitution industrialization articulating the different regions of the country anew. As sugar prices boomed, the dictator expropriated the country's sugar industries from their US owners and doubled the area under cultivation while tripling output (Turits 2003). Rural producers of the southeast were massively dispossessed, swelling the ranks of the urban poor and working classes in Santo Domingo. But dispossession in the Cibao was also gaining pace as the commercialization of tobacco and coffee on *minifundios* incorporated peasants into unequal market relations that led to increasing indebtedness and growing numbers of seasonal wage workers involved in regional circuits of labor migration to supplement declining farm incomes (San Miguel 1997; Lozano 2001). Parallel to the nationalization of the sugar industry, the regime turned from peasant production to import substitution in agriculture and industry. Trujillo promoted staple food production, especially rice, together with agro-processing in the Cibao valley, while stimulating industrial production of consumer goods in the capital city, activities that were protected behind high tariff barriers and subsidized by generous tax holidays (Moya Pons 1992).

In the post-dictatorship period following Trujillo's assassination in 1961, the slow turn away from smallholder production and the weakening of ISI promotion of domestic agriculture accelerated as state subsidized credit and non-pecuniary assistance for rural producers and agro-industry declined (Dore y Cabral 1981; Betances 1995). If Cibaeño peasants and small farmers responded to these shifts in large part through domestic and international emigration (see Chapter 4), the Cibaeño elites sought access to industrial incentives instituted by the Balaguer government in 1968 under Law 299. The incentives law ostensibly represented a compromise between Balaguer and his US backers who had supported his accession to power in 1966 following the US' second occupation of the country (see Hartlyn 1998; Schrank 2003). The law created two categories of industrial incentives: an ISI category that favored domestic industries through a set of subsidies and market protections, and an industrial export category which established tax and tariff exemptions for companies that produced for export in trade zones. In practice, throughout the 1970s, Balaguer resisted adjusting budget priorities to US dictates and orienting the Dominican economy towards manufactured exports. Indeed, ISI incentives were granted three times more frequently than those for export, and nearly 80 percent of ISI incentives were granted to companies located in the capital city, considered to be clients of the Balaguer regime (Moya Pons 1990; 1992).[28]

These dynamics mapped on to the differing agrarian structures of the country to lasting effect. The first two trade zones were established in the southeastern sugar area with the express purpose of absorbing male workers retrenched from the sugar mills as the sector restructured (Abreu & Cocco 1989; Schrank 2003).[29] The US-based transnational Gulf & Western, the new owner of the only foreign sugar enclave to survive Trujillo's nationalization, established the first trade zone in 1969 in the town of La Romana, close to its sugar installations. The second zone was founded three years later in nearby San Pedro Macorís with support from the state industrial promotion agency. While employment in both these zones would grow significantly once structural adjustment was implemented, levels of local ownership and capital investment would remain low. The principal investors would be of Taiwanese, Korean, and US provenance, seeking market access through the generous quotas of the Special Access Program (Schrank 2008). As competition stiffened in the late 1990s and market access was eroded, these firms would largely disinvest from the country and move their operations elsewhere.

Suffering from Balaguer's relative neglect of agriculture and agro-industry and the exclusion of Cibaeño firms from ISI incentives, Cibaeño regional elites established the country's third trade zone in the inland regional capital of Santiago in 1973 (Moya Pons 1992; Santana 1994).[30] The zone was owned and operated by a non-profit corporation, the Association for Development, Inc. (APEDI), established by the commercial elite in the 1960s to advance regional development projects. While foreign ownership was prevalent during the 1980s, local investors began to take over these operations or to establish new firms, while APEDI successfully lured not only foreign investors, but also return migrants from New York and Puerto Rico with experience in the garment industry (Schrank 2008). The Santiago trade zone served as a kind of incubator for a handful of Dominican managers and engineers, some from less established families, who would become large firm owners. Several of these managers created strategic partnerships with regional factions of capital and moneyed Dominicans in New York in order to start joint ventures or sole-owned firms in the 1980s when capital requirements were still relatively low (Guarnizo 1993, Schrank 2003). Already by 1986, the Santiago trade zone housed twice as many locally owned firms as the San Pedro and La Romana zones combined (Schrank 2008).

As I discussed in the previous section, the initial boom and spectacular growth of the Dominican garment sector began to wane in the mid-1990s as other producers from the circum-Caribbean and Mexico entered into competition for US market share. Predictions of the effects of intensified competition on Dominican garment producers were dire (Mortimore 1999; Matthews 2002), but the results of changing regulatory and competitive conditions were mediated by the particular trajectories of accumulation and disinvestment in the sub-regions of the country. While the total number of firms and workers indeed declined by the end of the decade, garment production consolidated in the northern Cibao region and employment there rebounded (Table 2.2).

Changing trade regulations following the passage of so-called NAFTA-parity for the circum-Caribbean permitted duty-free treatment for garments that underwent transformations such as industrial cutting and finishing, in addition to simple assembly, in the macro-region. A handful of Santiago-based firms began to incorporate these functions, together with other backward and forward linkages, which I detail in the following chapter. The transformation of these firms into higher "value-added" ventures involved the development of long-term sourcing relationships with US buyers who provided credit, technical expertise, and wider industry connections in exchange for low-cost finished goods and the ability to disinvest from the production process. In general, these lead contract suppliers, headquartered in Santiago, remained vertically integrated, incorporating new parts of the production process while still directly employing large numbers of sewing operators, in addition to creating their own subcontracting networks throughout the Cibao

**Table 2.2**  Dominican trade zone employment, 1992–2004

|  | 1992 | 1996 | 2000 | 2004 |
|---|---|---|---|---|
| Total garment employment | 100,437 | 107,867 | 141,945 | 131,978 |
| Percentage of trade zone total | 71 | 66 | 73 | 70 |
| Percentage of trade zone total in northern region | 42 | 43 | 51 | 52 |

Source: Annual reports, CNZFE, various years.

valley, and eventually, as I describe in Chapter 5, in Haiti. Indeed, by the beginning of the 2000s, three of these Santiago-based firms accounted for between one-fifth and one-quarter of the country's entire garment workforce (see Chapter 3). These producers and their allies formed a powerful political association that lobbied for US market access by promoting new generation trade agreements that were quickly replacing unilateral arrangements like the CBERA. Indeed, the garment sector-dominated trade zone lobby, which funded a permanent legal representative in Washington, was widely considered to be solely responsible for the Dominican Republic's adhesion to the Central American Free Trade Agreement in 2004 (see Werner & Bair 2009).[31] By the early 2000s, many of these entrepreneurs were diversifying their businesses by acquiring or starting export ventures in other sectors including call centers, ceramics, and medical devices.

Executives and owners that I interviewed in Santiago represented their relative success in manufacturing exports in terms of the Cibao region's particular "tobacco" culture. Their claims were also represented by tobacco-inspired art and maps of the Cibao that adorned the walls of waiting rooms and executive offices in the region's garment factories. The managing partner of IA Manufacturing, for example, himself a member of a prominent commercial family of Levantine origin in the region, linked the rise of the company, the second largest Santiago garment firm by employment at the time, to the tenacity of Santiago garment owners embedded in the Cibao's entrepreneurial culture:

> There were people who took risks. Lots of people didn't have any capital. [IA Manufacturing's owner] put his house up as collateral and started the factory with almost no capital. He started with 300 workers and now has 10,000. But all the money that came in went right back into the business ... In the Cibao, there is what has been called a culture of tobacco, the difference between minifundios and latifundios, a culture of saving and a culture of debt. In Moca [a nearby town where he first worked], everyone had a side business. Everyone. One woman lent money, another one sold merchandise, another one prepared food.

For the managing partner, reflecting the ideal of a cross-class common identity that I have discussed, the region's entrepreneurial culture extended from owners to workers who combined wage work with other market activities. This culture stood in contrast to that of the *latifundio,* which engendered profligate, unproductive spending, often implying moral and political corruption of both elites and workers. The owner of the Santiago zone's main logistics company, also from a prominent regional family, ascribed the success of the region's trade zone strategy directly to its distinction from the elite culture of politicking and corruption in the capital city:

> The initiative of the criollo sector here in different industries ... there is a different dynamic. The south [i.e., the capital] is dedicated more to making a living by politics. It's easier to dedicate yourself to politics than to put yourself to work in industry.

Obviously, such a statement minimized the sector's own investment in lobbying for market access in Washington. Nonetheless, what is evident from the Cibaeño elite's self-representation is that their efforts to carve out niches for profit-making within often unstable and highly competitive global production networks was inseparable from inter-regional struggles over the creation and distribution of surplus, and the class identities forged in and through these processes.

## Conclusion: Moving beyond "Global Forces" versus "Regional Paths" Approaches

The garments-as-globalization story is indeed a powerful, if inadequate, account of change in the circum-Caribbean in the 1980s. Industrial restructuring brought on by the advent of flexible production, Cold War geopolitics, and neoliberal trade reforms coalesced to produce a dramatic transformation in the export activities of circum-Caribbean countries over the course of the decade. And yet, the uneven trajectories of the subsequent garment, and wider export manufacturing, boom are often taken for granted by reframing these differences through comparative debates on the successes and failures of capitalist development. The global factory thus operates not only as a site of investment and exploitation, but also as a producer of meaning, one that frames manufacturing activity as a stage of development to be traversed or not by a given country. As we will see in Chapters 5 and 6, the erasure of uneven development and the persistence of such sequential narratives is part and parcel of reproducing "failed" places as available for later rounds of accumulation through similar policies.

The regional story of garment export growth and consolidation in the Dominican Republic offers an important corrective to the globalization account. Here, we see how regional trajectories of capitalist accumulation shape, and are shaped by, agrarian histories and rounds of investment and restructuring. To date, however, much that is written about this phenomenon – in the Dominican Republic and beyond – seeks to interrupt the spatially homogeneous globalization narrative through regional histories that presume local capital as the primary mover and determinant of place-based outcomes.[32] Moreover, these attempts almost invariably focus on success stories while failing to theorize how such places with developmental elites are reshaped by the recursive dynamics of disinvestment and devaluation that are constitutive of capitalism. As geographer Doreen Massey pointed out many years ago, regionally based capital is not necessarily regionally loyal (1995 [1984]). Certainly, garment jobs were more abundant, local ownership more prevalent, and capital intensification more pronounced in the trade zones of Santiago and the surrounding Cibao relative to other regions of the country during the period of the boom and subsequent consolidation. Yet, with further intensification of competition in the mid-2000s following the implementation of WTO trade rules, local garment owners took a variety of paths: some diversified into other activities, both productive and purely speculative, abandoning garment production entirely in many cases; others sought to create new outsourcing arrangements in Haiti and beyond; and others deployed some combination of these options.

In short, divergent regional outcomes are neither purely products of global capital's desires nor local elite projects. If we are to better understand the processes through which geographies of uneven development are made and remade, we must turn to a more dialectical analysis of class relations in articulation with shifting constructions of race and gender. We must begin by interrogating how a dominant accumulation strategy such as export manufacturing intersects with places that have long been shaped by particular forms of inclusion within – and exclusion from – global markets. And we must ask how subaltern livelihood strategies forged in and through these volatile geographies of accumulation contribute to the making of place. Thus, the aim of the chapters that follow is to disrupt accounts of export restructuring as either the product of global forces or of particular local paths. Excavating specific, yet linked, trajectories of uneven development between the Dominican Republic and Haiti, I offer a critique of dominant development narratives that have sought to reduce complex spatio-temporal processes into an ever repeating imperative of global market integration. In the wake of this critique of development, I focus on how regional dynamics of restructuring and disinvestment are shaped by historical trajectories and the everyday politics of value practiced by the people included within, and excluded from, these shifts.

## Notes

1   The creators of the program became key experts in export promotion strategies not only in the Caribbean but also in East Asia (see World Bank 1994). The Caribbean Basin Initiative, discussed below, was premised upon similar assumptions and incentives.

2 For 1980, Bulmer-Thomas lists manufactured exports as a percentage of total exports for key export-promoting Latin American countries as follows: Brazil (37.2 percent), Colombia (19.7 percent), Dominican Republic (23.6 percent), Haiti (58.6 percent), Mexico (31.1 percent) (Bulmer-Thomas 2003: 320). Haiti's high percentage may be an indicator more of dependence on manufactured exports than high output, perhaps related to a decline of Haitian agricultural exports.

3 Eight days after taking office, President Reagan wrote the following in his diary: "Visit by P.M. Seaga of Jamaica, his wife and members of his admin. Our 1st state luncheon. He won a terrific election victory over a Cuban backed pro-communist. I think we can help him and gradually take back the Caribbean which was becoming a 'Red' lake" (quoted in Klak 2009: 18). In addition to the clear concern for the spread of Communism, the metaphor of a lake evokes a similar idea of proximity and control as the more familiar notion of the region as the United States' backyard.

4 Grandin (2006) argues that Reaganite policies drew upon the ideological legacies of US imperialism in the circum-Caribbean in the nineteenth century. The most relevant ideologue from this era, Alfred T. Mahan, theorized the Caribbean in terms not dissimilar to the geopolitical imaginary that underpinned the CBI. See Rodríguez Diaz (1999) for a comprehensive review of Mahan's geopolitical imaginary of the Caribbean.

5 Pantojas-Garcia (1985; 1990) argues that the economic component of CBI had many similarities with Operation Bootstrap. US policymakers intentionally downplayed these parallels, however, because of the abysmal condition of Puerto Rico's economy at the time that CBI was being debated.

6 US sugar import quotas placed on the Caribbean were reduced in tandem with increased domestic sugar production (beet and cane). Between 1981 and 1987, US imports of circum-Caribbean sugar decreased from 1.6 to 0.3 million tons. Sugar consumption in the United States also declined due to the proliferation of sugar substitutes, especially high fructose corn syrup and synthetic sweeteners, while US domestic sugar production increased by 15 percent between 1981 and 1989 (Messina & Seale 1993).

7 Because the exemption from paying duty is not reciprocal – that is, initially circum-Caribbean countries were not obligated to drop tariffs on imports of like goods from the United States – the CBERA is considered a preferential agreement. Given the restriction of the duty-free status to non-competing goods and the initial exception in place for garment and textile, however, the term "preferential," while technically correct, is also misleading.

8 Most of the goods granted duty-free access under the program were already guaranteed similar treatment under the multilateral General System of Preferences, the earlier package of postcolonial trade preferences established to grant a modest amount of market access for global South exporters to rich country markets.

9 In the early 1980s, about two-thirds of garment firms in the Dominican Republic and Puerto Rico were directly owned US subsidiaries (Safa 1995b: 11). In both cases, local capital ownership increased over time. Haiti appears to be an exception to this trend. Already in the 1970s, 71 percent of firms were

owned by Haitian capital. Quota levels were reduced on East Asian producers in 1984. As a result, Korean and Taiwanese firms moved operations to the region, increasing their share of plant ownership. One report found that 23 Taiwanese or Korean garment assembly firms established operations in the CBI region in the two years following the quota reduction (Rosen 2002: 147).

10   In turn, new provisions of the Multifibre Arrangement created special quota categories for OPT production, which permitted this macro-regional arrangement to be in compliance with the multilateral trade system (i.e., the GATT).

11   The phrase has various attributions. Secretary of the Treasury Don Regan argued for the need to Taiwanize Mexico during that country's default in 1982 (Menéndez 1982; also Fernández-Kelly 1992). In testimony before Congress that same year in support of funding for export promotion in Haiti, a United States Agency for International Development (USAID) official, M. Peter McPherson, argued that closer ties to the United States would "make the prospects for Haiti as the 'Taiwan of the Caribbean' real indeed" (quoted in DeWind & Kinley 1988: 61). According to DeWind and Kinley, the phrase was widely used by promoters of export-led development strategies at the time.

12   There is a large literature dedicated to addressing the question of why East Asia overtook Latin America in terms of income per capita and industrial growth by the end of the 1980s. Classic contributions include Gereffi and Wyman (1990), Evans (1987; 1995), and Hein (1992). This literature focuses on the differing institutional contexts, the role of the developmental state, and the action of exporting firms that learned from their customers – that is, multinationals – to later become their direct competitors.

13   In its famed East Asian Miracle report, the World Bank was at pains to portray the growth of East Asian economies as the result of market fundamentalist policies similar to those that the international financial institutions (IFIs) advanced under structural adjustment. The Bank faced an internal rebellion from Japanese representatives and much criticism from "developmental state" scholars who emphasized the major role played by the state in protecting industries, restricting outward capital flows, and, in the case of Korea especially, restricting foreign direct investment (Amsden 1994; Wade 1996; Chang 2006).

14   Preferential trade arrangements, limited as they were, recognized asymmetry between trading nations, particularly between former colonies and their colonial or neo-colonial metropoles (see Chang 2007; Girvan 2010). With the creation of the WTO, and the demise of the Communist bloc as a viable competitor for Third World allegiances, this postcolonial principal together with the preferential trade arrangements that enshrined the principal were largely abandoned. Structurally disadvantaged trading countries must adopt the same rules as those countries that sit at the top of the global trade hierarchy, only occasionally given more time to implement the rules (Chang 2007).

15   At least 82 more Korean and Taiwanese firms established operations in the region between 1989 and 1991 (Rosen 2002; see also note 9).

16   The austerity measures were announced the day before the Easter holiday (Semana Santa). When stores re-opened, prices had risen by approximately 30 to 50 percent or more for basic foodstuffs (Artiles-Gil 2002) The *New York Times* (1984) reported that the price of a pound of beans had risen from the equivalent of 0.30 to 0.65 dollars and cooking oil from 9.75 to 25 dollars a gallon.

17   For a political analysis of *la poblada*, including the role of neighborhood associations and the Communist Party, and the lasting effects on popular movement organizing, see Artiles-Gil (2002).

18   Quoted in the *New York Times* (1984). Blanco returned from Washington empty-handed. Although a significant recipient of aid in the 1960s, the Dominican Republic received substantially less aid than its circum-Caribbean neighbors that were on the frontlines of the Cold War during the 1970s and 1980s. *La poblada* does appear to have eventually spurred additional donor dollars since the United States was loath to lose a "model" client in the region (see Meislin 1984; Artiles-Gil 2002).

19   Various observers of this phenomenon describe the problem as a "fallacy of composition." The same problem was evident in the competition for primary commodity markets wherein multiple countries would promote a particular commodity only to face increased competition from other countries undertaking the same strategy, thus driving down their returns (e.g., Kaplinsky 1993; Heron 2006). The dynamic is even more aptly described by Schumpeter when he writes that firm strategies can only be analyzed as part of capitalism's overall organic process of seeking profitability: "Every piece of business strategy acquires its true significance only against the background of that process and within the situation created by it [as part of] the perennial gale of creative destruction" (2008: 83–84). Although Schumpeter describes creative destruction here in terms of an organic analysis of business strategies, the same could be said of regulatory dynamics.

20   Kaplinsky (1993) points out that because the US dollar was devalued over the latter half of the decade, circum-Caribbean currencies experienced even stronger devaluations relative to a basket of world currencies.

21   Recall that under the so-called production-sharing 807 model that predominated in the circum-Caribbean, a tariff was levied on the amount of value added – that is, labor – to the product.

22   The highly politicized debate surrounding the collateral impacts of NAFTA on circum-Caribbean garment producers led to the passage of a weak form of "NAFTA-parity" in 2000, dubbed the Caribbean Basin Trade Partnership Act (CBTPA). The new legislation extended duty-free treatment to "local value added" – that is, labor – but failed to allow circum-Caribbean producers to use textiles from Mexican, Canadian, or circum-Caribbean producers (Heron 2004).

23   The Dominican Republic was sparsely populated and weakly integrated into regional and global markets in comparison to neighboring Haiti, the

result of the territory's relative abandonment by the Spanish crown for much of the colonial period, in contrast to French Saint-Domingue, which was the center of the French colonial empire during the eighteenth century. By the 1920s, a mere 30,000 people lived in Santo Domingo, the first city established by Cristobal Colón in the Americas four centuries earlier; just over 15,000 people resided in Santiago and agriculture covered a small fraction of the country's landmass (Santana 1994: 67; Turits 2003: 16).

24  Pedro Bonó, a middle-class Cibaeño lawyer regarded as the Dominican Republic's first sociologist, was also the first intellectual figure to critique the prevailing ideology of progress associated with the country's growing sugar frontier and to advance the small farmer as an alternative model associated with tobacco cultivation, a crop he called the "true father of the fatherland" (quoted in San Miguel 2005: 14). In 1884, Bonó wrote: "I have seen the transformation of the East; property transferred almost free of charge to new occupants concealing themselves under the guise of Progress. Progress it would be if what was occurring was progress for Dominicans, if the old peasantry ... were in part the owners of the sugar mills and plantations ... But instead of that, while before they were poor and rough, at least they were owners, now they are coarser and poorer proletarians. What form of progress is that?" (quoted in Turits 2003: 63). Bonó's ideas would gain traction and take a nationalist turn following the 1898 Spanish–American war as US capital came to dominate more of the country. By the 1920s, under the weight of US occupation on both sides of the border, Dominican elites elevated the *criollo* farmer as a national symbol, the pillar of "an imagined small-farmer mode of development" (Turits 2003: 65).

25  For Bonó, on the other hand, all Dominican peasants were the "offspring" of slavery mixed with the descendants of poor colonists (Turits 2003: 25–26). According to San Miguel (2005: 45–58), Bonó was somewhat exceptional for this "mulattoism," an ideological current that, on the one hand, valorized race-mixing and sought to ground Dominican identity in a notion of racial harmony, while, on the other hand, desired a "dilution" of blackness as part of elevating the market-integrated Dominican peasant as a national symbol. In contrast, as I discuss below, twentieth-century anti-Haitian ideologues would deny the mixed race character of the peasantry altogether.

26  In this sense, drawing on currents that stemmed from Bonó prior to the rise of Trujillo, *dominicanidad* – as read in and through the Cibao region – was an idealized vision of both class cooperation and racial harmony.

27  The massacre targeted independent producers. It did not interrupt the Haitian migrant labor stream. Haitian sugar workers continued to labor on the southeastern plantations throughout the massacre and its aftermath. For a powerful literary account of the massacre in English, see Danticat (1998).

28  Six hundred and ten of the 773 industries granted incentives under the import substitution category of Law 299 between 1968 and 1982 were located in Santo Domingo. Over the same period of time, only 240 businesses received approval for export incentives (Moya Pons 1992: 420).

29 Similar to the initial logic behind the Border Industrialization Program in Mexico, which was established ostensibly to absorb unemployed, predominantly male, Mexican migrant farmworkers following the termination of the United States' *bracero* program, these new zones, like Mexico's *maquilas*, primarily recruited their wives and daughters instead (Fernández-Kelly 1983; Safa 1995b).

30 Trade zones in the capital city would not be established until the end of the 1980s, when *capitaleño* investors began to switch their industrial strategy away from ISI (Abreu & Cocco 1989: 68). These zones in the Santo Domingo area would generate less employment, be dominated by multinational branch plants, and remain largely assembly operations (Schrank 2008).

31 For a complementary perspective on the importance of export-oriented business networks formed during the CBI period to the passage of DR-CAFTA, see Cox (2008).

32 See Gertler (1992) for an early, comprehensive critique of these tendencies in the so-called new regions literature of the 1980s, represented by iconic models of success in the face of global competition, such as the Third Italy. See Hadjimichalis and Hudson (2013) for a recent critique of the depoliticization of regional theory and the sidelining of questions of uneven development, which they view as a theoretical consequence of the neoliberal policy turn in the 1990s.

# 3

# From *Manufactura* to *Mentefactura*?

## *Gender and Industrial Restructuring in the Dominican Republic*

### Introduction: Beyond "Cheap Labor" in the Dominican Republic?

In the hollowness of a shuttered factory in the Santiago trade zone, the owner and founder of the largest garment firm in the Dominican Republic, Dominican Textile (DT), addressed a forum of export industrialists and their allies in government, flanked by a small crowd of about 100 hand-picked managers and the press. Standing before the audience, the owner held up an industrial spool of thread and a printed t-shirt and declared the following:

> I created 14,000 jobs. I have 11,000 jobs left and we are all fighting for those 11,000 jobs and maybe a few more ... We have to continue to improve our internal processes and continue to convince from the President of the Republic to many government bureaucrats to our legislators that we are not a footloose company. We are not a low-tech company as some state officials who want to deride the sector make us out to be. We are not a company that can exchange sewing machines for computers from one day to the next. We have a population of employees from the age of twenty-five on who will neither become efficient with computers nor will they learn perfect English to work at a call center ... For those who say that we are, are the past, there are companies who are buying this [thread] in the world market to make a product like this [t-shirt]. A *muchacha* [young woman] from Gurabo designed this. Another *muchacho* [young guy] sells it in New York, goes to New York and sells

*Global Displacements: The Making of Uneven Development in the Caribbean*, First Edition.
Marion Werner.
© 2016 John Wiley & Sons, Ltd. Published 2016 by John Wiley & Sons, Ltd.

the product ... This one is Donna Karan. This polo shirt, we did the cloth, the weaving, the dyeing, the marking, the cutting – the sewing – and this, yes, you could say is low technology ... but the girl who designed it, the guy who marketed it, the one who printed the design, designed the components which were all bought locally – the labels, packaging. This is not low technology. When you hear people of authority saying that we are going to disappear because we don't have technology, it's simply not true. This [t-shirt] is high technology.[1]

As the sector faced a profit crisis, the company owner offered the high-tech t-shirt as the solution for the industry and, by extension, for the region. The owner advanced a vision of an industry that would sustain low-wage assembly jobs while signifying the sector's value through the work of its cosmopolitan employees: the young woman designer from Gurabo – a town-turned-Santiago-suburb now host to the trade zone model's winners on its prestigious hillsides and working-class rural migrants on its margins; and her co-worker, a jet-setting transnational vendor who moves fluidly between Santiago and the pinnacle of the global apparel industry, New York City.

In its intensive lobbying activities, the garment sector was defending itself against an apparently divergent vision of a high-tech path for the country advanced by government officials and international development policymakers, one that excluded garment production altogether. For these actors, the imperative of progress could only be achieved if Dominican firms moved towards new sectors like call centers and information technology services. In the government's high-profile National Competitiveness Plan, for example, garment and textile production was presented as a stage that would have to be overcome in an inexorable march toward global competitiveness. The authors of the plan asserted that as one of the "countries of 'late industrialization,' [the DR] cannot commit the mistake of arriving late again to the era of knowledge" (CNC 2005: 1). The country's competitiveness in the US market would depend on export industries making "the successful transition from assembly (first generation) to value-added manufacturers (second generation) ... to *mentefactura* or *menteobra* [mind-facturing or mind-labor]" (CNC 2005: 31).

In this chapter, I show that this apparent conflict between garment owners and policymakers relied upon gendered structures of meaning as key mechanisms to reproduce the conditions of accumulation for the country's export sector.[2] Both the notion of the high-tech t-shirt and the idea of a transition to "*mentefactura*" rested upon resignifying the low-wage, so-called unskilled feminized worker of the assembly model from a symbol of development and progress, to one of stagnation and decline. I develop this argument by demonstrating the parallels between the ways that gender was mobilized to create "value-added" production in the Dominican garment sector, on the one hand, and World Bank assessments of the Dominican export sector wedded to gendered economic dualisms of non/value, on the other.

My argument draws upon and develops a feminist analysis of the dialectic of labor as an embodied subject and as the abstract form of value. At the center of these interrogations is a theoretical approach to the politics of capitalist value that focuses upon the practices and discourses that transform the variable content of labor – that is, embodied, living labor – into the abstract form of value (Elson 1979).[3] In research on the global factory, feminist scholarship has interrogated this dialectic through the study of the ways that gendered practices and discourses produce the feminized subjects that occupy the lower rungs of the global production hierarchy (e.g., Wright 2006; Freeman 2000; Collins 2003; Ramamurthy 2010; Salzinger 2003). Feminist work in this tradition asks *whose* labor is constructed as valuable and *how* this construction is achieved. Eschewing fixed or essential characteristics of workers and skills, feminist scholars have argued that gendered labor is not simply hired; it must be produced by managers, owners, workers, households, and the state through discourses and practices that shape expectations and assumptions about what kind of labor is valued in production, and what kinds of bodies are suitable for that work. With respect to gender in particular, feminine and masculine norms and signs operate relationally: rather than denote fixed categories of work, we will see how gender categories serve as productive technologies that construct certain kinds of work as valuable or valueless, on the one hand, and associate certain jobs with certain kinds of bodies, on the other hand.

My study of garment sector restructuring demonstrates that gender is not solely germane to the construction of feminized labor at the lower value rungs of global production networks, however. Rather, gendered meanings and practices are just as salient to the restructuring of those networks. In fact, gender norms serve as a kind of resource to rework the hierarchical field of positions in global production networks, while remaining largely unexamined and unquestioned because they often appear as either natural or outside the proper domain of capitalist production. In short, the struggles over the meanings and practices of export restructuring that I describe here build upon the two central observations of feminist analysis: first, that the heterogeneity of labor lends flexibility to the labor process; and second, that subjects are produced together with labor process transformations. Feminist analysis can thus shed light on the gambit of gendered practices and discourses that are mobilized to renew the conditions of accumulation in restructuring global networks of production. Finally, I extend a gender analysis from flexible production to flexible regulation through an analysis of the development discourses associated with export restructuring. Similar to Quijano's observation of Eurocentrism as a paradoxical amalgam of stage-ism and dualism (see Chapter 1), in these narratives, gendered structures of meaning reproduced and resignified economic dualism in order to recuperate stage-like development in the face of neoliberal failures.

I elaborate my argument in three subsequent sections. The following section describes the geographical and organizational restructuring of the garment production network that spans the United States and the Dominican Republic, and the regional restructuring of firm networks and capacities in the Cibao. These changes were associated with increasing participation of male workers in firms that had incorporated "value-added" processes; I interrogate predominant explanations of this "defeminization" of the garment workforce. In the third section, "Gender and the Reworking of Sewing and Skill," I use a feminist analysis to interpret these trends through a detailed case study of Dominican Textile (DT), the country's largest apparel company in terms of investment and employment. I focus upon the narratives and practices surrounding the gendering of skill in two areas, sewing and engineering. Drawing on in-depth interviews and observations in the firm, I show how gendered meanings of skill and practices of training worked together to construct skilled sewing as a masculine activity, rationalizing the progressive exclusion of female sewers from the upper rungs of a new sewing skill hierarchy. In contrast, I discuss the creation of a new occupation at the firm, the service engineer, whose role was to translate the brands' and retailers' demands into the production process. I discuss the gendered discourses and class position that shaped understandings of this occupation, and the almost exclusive employment of women in this job. The fourth section, "Gendered Economic Dualisms," analyzes the World Bank's narrative of Dominican export restructuring from the 1990s, at the height of the free trade zone (FTZ) boom, to the mid-2000s, as the low-wage export model faltered. I focus upon the portrayal of development through the Bank's notion of the Dominican Republic as a dual economy, comprised of the trade zone and the "non-trade zone" sectors. In dialogue with the production network restructuring process at the heart of the chapter, I explore the analogous and supportive sign-formations of market/non-market, modern/traditional, mind/body, and male/female that are drawn upon to signal progress to a new "stage" of Dominican development by naturalizing the expulsion of feminized labor.

## Cibaeño Firm Restructuring and New Production Geographies

Dominican producers were part of a garment production network organized into five stages along a very specific transnational pattern until the 1990s. The process commenced with brands, retailers, and manufacturers located in old or emerging fashion centers of the industry in the United States. These firms created or commissioned designs, translated designs into industrial patterns, graded the patterns for size, and selected and sourced cloth and components. The second stage was the production of cloth, which was either knit or woven and dyed in textile mills in the US south. The third stage consisted of preparation for assembly in Tampa or Miami, Florida. Preparation for assembly included cutting, sorting, and bundling fabric pieces. The fourth stage was assembly, which took place in the foreign and domestically owned firms in the Dominican Republic. The final stage was finishing, which included post-production treatments of cloth such as grinding and sanding of jeans, industrial laundering, pressing, and packaging. The same firms in Florida carried out finishing, while brands, retailers, and manufacturers coordinated other post-production functions including marketing, quality control, and later, auditing for social standards.

Each of these five stages was linked through a shifting mix of direct ownership and outsourcing among firms. While most businesses performing the functions at each stage of the production process were formally independent, as development sociologists have long demonstrated, the labor process, or commodity chain, was "driven" by the industry's lead buyers – that is, the brands and retailers based in the fashion centers of the industry (Gereffi & Korzeniewicz 1994; see Chapter 1). These buyers coordinated and controlled the process, and set the terms for the participation of suppliers in the production network. Manufacturers, similarly, outsourced to the region and some, over time, divested entirely from their US-based manufacturing facilities, becoming brands and, to a lesser extent, retailers. The San Francisco-based company Levi's is one particularly relevant example: the firm established long-term relationships with key Dominican suppliers to produce casual pants and jeans for the US market as it divested from its wholly owned manufacturing plants in the United States, shutting down its remaining production facilities in 2003 (Dicken 2011).

In the Dominican Republic, the restructuring of firms was prompted by changes in regional and global trade rules and the ambitions of a local capitalist elite (see Chapter 2), together with the changing strategies of lead retailers, brands, and manufacturers in the United States. The resulting distribution of functional capacities was highly uneven across the sector and the country. Many firms continued as assembly-only operations, while others incorporated a reduced number of functions for cutting and simple finishes in addition to sewing (called CMT, or cut-make-trim). Meanwhile, a handful of assembly firms developed the capacity to carry out or to coordinate more parts of the labor process. In the literature, these firms are called "full-package" suppliers. A brand or retailer could outsource nearly the entire process to one of these suppliers, divesting from the risks and costs associated with any part of production. In fact, Levi's was instrumental in the transformation of four assembly suppliers into full-package producers in the Cibao. As part of the company's divestment from production in the United States, the manufacturer-turned-brand provided credit, technology, and other forms of expertise to these Cibao-based companies. These firms eventually emerged as the principal full-package operations in the country. They repaid the brand with low-cost, finished apparel, an arrangement that significantly reduced the capital expense to create these new functional capabilities. Moreover, the Dominican suppliers hired on high-skilled experts that the brands recommended and trusted.[4]

The emergence of full-package suppliers transformed the structure of the local industry, establishing a new hierarchical arrangement. The Dominican-owned full-package enterprises came to occupy key supplier positions, working directly for US brands and retailers, and subcontracting work to assembly and CMT firms locally, while still retaining, and even increasing, their own direct employment in sewing. Nearly all these Dominican lead enterprises were based in the country's northern region, consolidating Santiago and the surrounding area as the definitive center of the garment trade. The three largest full-package firms employed 5–14,000 workers each, accounting for between one-fifth and one-quarter of the total export garment workforce, not including the indirect employment these firms controlled locally through their large subcontracting networks.[5]

Cibao-based full-package firms restructured primarily through vertical integration, establishing wholly owned or joint-venture supply companies, as well as in-house capabilities in design, sample and pattern making, cutting, laundering, and finishing. These transformations reshaped the industry's spatial pattern. Responding primarily to changes in trade rules which no longer required garments to be cut in the United States in order to receive tariff free re-import to the United States, Dominican-based suppliers purchased cutting rooms and associated functions and eventually moved the equipment from Florida to Santiago.[6] To make such investments, and to access credit lines in order to finance the cost of sourcing textiles, by far the most expensive input in garments, the emerging full-package firms created alliances with Dominican investors from the country's traditional industrial and financial groups.[7] DT was the most aggressive in developing its full-package capabilities. The company formed a joint venture with the US-based A&E company to set up a thread supplier serving much of the cluster. DT also diversified from woven garments – principally jeans and casual pants, which were the mainstays of the sector due to its strong alliance with Levi's – to knit garments. The Dominican Republic had to import all woven fabric, mostly from the United States due to trade restrictions, because the country had no woven textile mills. DT invested in the much less capital-intensive process of manufacturing knit material by purchasing a knit mill in Alabama and transferring the machines to the Dominican Republic in 2003, thereby boosting the country's limited capacity in knit textile production and establishing only the second domestically owned industrial-sized knit plant in the country.

Restructuring at DT and other Dominican firms was on-going as firms responded not only to changes in buyers' demands for supplier capabilities, but also to additional changes in international trade regulations into the 2000s. With the final phase-out of global quotas on apparel at the end of 2004, large continuous orders shifted toward lower-wage producers in Asia (see Chapter 2). The result of the end of the global quota system was two-fold. First, many Dominican-owned and most foreign-owned assembly operations shut down: the number of garment factories decreased from 281 to 143 between 2004 and 2008. Combined with downsizing in surviving firms, the industry retrenched 62 percent of its workforce between 2004 and 2008, falling to 49,735 workers. By 2012, the number of garment workers in the country's trade zones stood at 40,666, or about one-third of the number before the quota phase-out (CNZFE 2005; 2009; 2013; see Figure 3.1).

Second, on the production floor, Dominican firms faced pressures from their US buyers to transform the organization of the labor process in order to satisfy the demand for low-volume orders and an increasing number of styles. Some of these orders were to replenish shelves in North American stores in the middle of fashion seasons, viewed as a possible future for the sector, given the country's geographic proximity to the United States (Abernathy et al. 2006). Thus Dominican firms had to adjust to a higher volatility in demand, shorter production runs, and a higher rate of changes in styles. In the case of DT, large, continuous orders of blue jeans were moved to the firm's new trade zone on the Haitian–Dominican border in 2003, where the per minute assembly rate was 35 percent lower than in the Dominican Republic – that is, 4.5–5 cents versus 7–7.5 cents respectively. Meanwhile, extra capacity for blue jeans, as well as for casual pants, tops, replenishment work, preproduction, and finishing, was handled

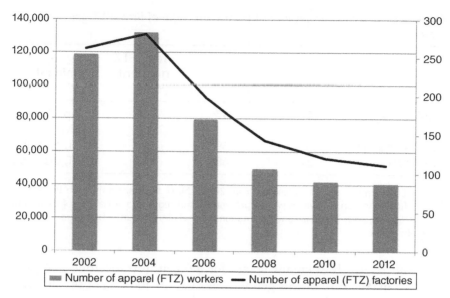

**Figure 3.1** Number of FTZ garment workers and factories, 2002–2012.
Source: CNZFE, various years.

in the company's directly owned facilities, or via its subcontracting network, in Santiago and the surrounding region.

While I take up the question of workers' experiences navigating the industry's employment collapse in the following chapter, here I lay out the data on changing gender ratios in surviving firms in order to unpack and explain these trends through a feminist analysis of labor process changes in the following section. Men's participation in the garment sector increased steadily over the 1990s and 2000s, reaching 45 percent by 2007 (CNZFE, various years). Studies of this trend in other countries have found that declining wages in other sectors, often combined with increasing or steady wages in trade zones, effectively pull male workers into trade zones as wages converge (see Barrientos et al. 2004 for a review). Indeed, in the Dominican Republic, wages for male trade zone workers began to converge with average national wages as the latter suffered stagnation and indeed decline in real terms in the late 1990s and early 2000s, and as trade zone wages for men increased (Kaplinsky 2000; PNUD 2005). Female workers did not see the same convergence because their trade zone wages on average remained below that of their male counterparts.[8] The eroding labor market position of Dominican men, combined with moderate increases in trade zone wages, no doubt contributed to men's increased employment in the garment sector. This explanation, however, accepts rather than explains increased wages for

**Table 3.1**   Regional employment by sex and type of firm, FTZ garment sector

| Region | Male (%) | Female (%) | Total FTZ garment (%) | Dominant type of firm |
|---|---|---|---|---|
| Northern region | 53 | 47 | 57 | Full-package, CMT |
| National district | 31 | 69 | 22 | CMT, assembly |
| Eastern region | 32 | 68 | 14 | CMT, assembly |

Source: Employment figures calculated from 2008 CNZFE figures, on file with the author. Type of firm based on interviews with representatives of regional associations of trade zone firms in Santiago, La Romana, and the national association in Santo Domingo, October–November 2006.

men and relatively stagnant wages for women in trade zones.[9] We can begin to make sense of this gender wage gap by looking at the different gender ratios found in different types of firms.

Indeed, variation in men's participation at the subnational regional scale provides strong evidence that the type of firm (i.e., assembly, CMT, or full-package) was an important determinant of male participation. In Santiago and the surrounding northern region, the area where the bulk of the country's full-package operations was located, men constituted 53 percent of the garment workforce (see Table 3.1). By comparison, men's participation in the other major garment clusters in the national district (i.e., Santo Domingo and the surrounding province) and in the erstwhile sugar-producing areas of the east was 31 and 32 percent respectively. Garment production in these two regions was almost entirely CMT and assembly.

The two regions that demonstrate similar rates of men's participation – the national district and the eastern region – have different labor market contexts (see PNUD 2008), indicating that labor market dynamics cannot explain the trend. The pattern across subnational regions suggests that the difference in gender participation rates is attributable to new, and better remunerated, employment opportunities available for men in full-package companies. This variation leads one to ask: why and how did men become the preferred workers in full-package firms?

When asked about the high participation rates of men in their operations, managers and owners in Santiago offered two explanations: local ownership and cluster specialization in woven garments. First, owners and managers cited the prevalence of Dominican-owned firms in the northern region, arguing that only foreign firms, which predominated in the other two regions, sought female workers. The minority partner of a large Dominican full-package supplier in Santiago offered a typical response when I asked if there had been a reduction in the proportion of women in the workforce. "Here [in the Cibao]," he explained, "there has always been, except for US companies that looked for female labor, here it has always been between 35 and 50 percent men in the companies owned locally, not the international ones." An increase in Dominican ownership of supplier firms in the 1990s in the region does appear to have coincided with the growing incorporation of men into the garment sector. And by 2004, the percentage of domestic firm ownership in the north was three times that of the east and nearly four times that of the capital (Schrank 2008). As I discuss in the previous chapter, the particular regional class dynamic of the Cibao – attributable to a history of more distributive land tenancy and a strong regional bourgeoisie – surely helps to explain the higher rates of Dominican ownership of garment firms in the northern region, as compared to the national district and the east. Whether we can attribute the increased proportion of male garment workers to the provenance of firm owners is, nonetheless, questionable. Certainly, regional elites were more adept at forging cross-class alliances around a strong regional identity, and drew upon more paternalist labor relations, thus relying less than foreign firms on predominant gender norms and assumptions about "docile labor." Evidence of these divergent relations of production can be found in the country's unionization patterns. Although low throughout the sector, nine collective bargaining agreements were in place by the early 2000s in export garment firms in the capital city and the east, where female workers predominated, giving short shrift to the assumptions of docile, female labor. In contrast, in the north, where male workers constituted nearly half the industry by that time, no such agreements existed, in part because of a paternalist approach to relations of production associated with regional capital.

The relationship between domestic ownership and the gender composition of the workforce, however, cannot be interrogated solely on these grounds, for an implicit gendered value of labor is inherent in the minority partner's response and remains unexplained. As we saw in the argument for the high-tech t-shirt, Dominican trade zone firms were at pains to distance themselves from the perception that they were "footloose" or "low-tech" companies. By associating a majority female workforce with foreign firms, Dominican managers and owners were both invoking a gendered norm – that is, women workers as devalued labor – and employing this norm to mark their own firms as "more advanced." What remains unexplained is how the norm itself was produced, and how shifts in the gender composition of the workforce intersected with the sector's restructuring as full-package firms in the Cibao attempted to consolidate their positions as higher "value-added" producers. I explore this question in the following section.

Second, several managers argued that the region's specialization in woven versus knit material offered another explanation for the preponderance of men in the Cibao garment industry (see also Safa 2002; USAID 2007). Indeed, in the mid-1990s, the Santiago-based cluster began to specialize in bottom garments (i.e., jeans and casual pants) made from woven material, a heavier fabric relative to the knit material used for top garments, and thus, these sources argue, more apt for male workers. While I do not simply dismiss this claim, it likely obscures more than it reveals. In the blue jeans cluster based in Torreón, Mexico, women made up the bulk of the workforce in assembly-only plants making blue jeans – that is, working with woven fabric. The proportion of female workers declined, however, as firms transformed themselves into full-package operations (Bair & Gereffi 2001; Bair & Werner 2011). The comparison between Santiago and Torreón strongly suggests that feminine and masculine labor are constructed in relation to one another, rather than to a particular product or material, and shaped through local labor market dynamics as well as labor process organization as we will see.

One final observation explains absolute increases in the number of men in full-package firms. Many of the labor process functions that were transferred from the United States to full-package firms in Santiago were gendered as male, and thus contributed to an absolute increase in male workers. In other words, the gendering of this work as masculine traveled with these functions and led to the almost exclusive employment of men within these departments.[10] Finishing, for example, employed many more men than women: the work of baking, washing, pressing, ironing, blowing, sanding, and destroying pants and jeans was low-paid and physically demanding. Barely mechanized, heavy physical work and the use of industrial chemicals contributed to the gendering of this labor as male. Moving cutting rooms, industrial laundries, and a textile mill corresponded to efforts by Dominican firms and their US clients to lower costs by employing workers, in this case male workers, who were devalued by the structural position of Dominican labor in relation to their counterparts in Alabama and Florida. Nevertheless, the absolute increase in male workers through the addition of these new departments (e.g., approximately 10 percent of employees at DT) cannot account for the overall trend of higher proportions of male employment in full-package versus assembly-only garment firms.

Thus, to deepen the analysis of how firms reworked labor's value in their efforts to occupy more profitable positions in the garment production network, we must engage in a more in-depth analysis of labor process transformations, changes that materialized through gendered narratives of skill and practices of training. In the following section, I discuss the gendering of sewing skills as masculine and contrast this trend with the gendering of engineering as feminine in the new client services department at DT; the comparison brings both gender and class difference into sharp relief. Taken together, my account demonstrates the flexibility of meanings facilitated by gendered norms and constituted through articulation, enabling the reconfiguration of the firm's exploitable workforce.

## Gender and the Reworking of Sewing and Skill

Prior to the company's transformation in the late 1990s, flexible labor at DT, like other assembly-only firms, was based on the progressive bundle system. Supervisors assigned individual operators single tasks, and skill was measured and rewarded on the basis of the speed at which operators performed their single operation (see Collins 2003). As DT sought to rationalize production to meet the demand for smaller orders and a larger number of styles, and to serve as the key node linking less capable local suppliers and foreign buyers, this paradigm of flexibility changed and new skilled sewing positions were created.

To produce for replenishment orders, managers required what they called multifunctional operators, combining the number of operations performed by a single operator. The company's wovens production manager explained that the new system was necessary to adjust to smaller replenishment orders, ranging from 1,500 to 5,000 pieces per week and short runs of rarely more than eight weeks, a pattern of demand that had become common since 2005. To keep a production unit, or module, working for an entire week and not to change styles midweek (affecting motivation, since workers would be less likely to make their production targets), the company sought to lower daily quotas in order to fill orders with fewer workers.[11] Accordingly, DT implemented a 15-grade pay scale that was indexed to the number of operations that workers undertook and the difficulty of the operations; wages ranged from 1,050 pesos to 1,950 pesos per week.[12] The job of plant engineers was to combine operations, assign an amount of time for the combination in units called "standard allowable minutes," and match these combinations to operators. Thus a reduced total number of operators completed lower daily quotas that were based on multiple operations over a four- to five-day period. To climb the wage scale, an operator had to demonstrate her or his ability to produce at 80 percent of the quota for additional operations on top of her or his base operation, to be completed at 100 percent. Every production line had at least one operator at the top grade; this was the utility position, an operator who could perform all the operations.

Once the engineer combined and distributed the operations and allotted time, each operator was expected to comply and faced a penalty if he or she refused to perform the multiple operations that were assigned.

Although data are not available on the precise number of men and women at each pay grade at DT, men clearly dominated the positions that were considered to be the most skilled, as judged by the difficulty of the operation or acquired skill in multiple operations. For example, male workers were dominant in the operations of sewing the inseam and attaching waistbands (see also Bair and Gereffi 2001). Men were perceived to be the most appropriate for these jobs because of the thickness of the woven material (see my earlier discussion); these were also the highest paid single operations according to the company's records. Moreover, utilities – the workers who could do all the operations – were exclusively male. Watching a plant engineer organize the operations on his spreadsheet for an upcoming production run, I asked how workers obtained skills in multiple operations. He explained that with the downturn in garment employment, the company could simply hire the workers with the requisite skills. Not only had no new formal training programs been implemented with the change to the new production system, but also the company had almost entirely scaled back on sewing training.[13] Barring formal training, informal forms of self-training constituted the main route for the acquisition of new operations.

To gain an understanding of how workers managed to train themselves and the gendered implications of this process, I spent time in the firm's sample room, part of its new product development department, where the firm's most skilled sewers worked. The process by which workers entered the sample room not only reveals the methods that the workers used to train themselves, but also sheds light on the role workers played in the strategies of firms to reposition themselves in the production network. Full-package firms like DT required sample rooms, part of larger product development departments, in order to work as direct contractors for US buyers. In fact, sample rooms and product development departments were repeatedly cited as the second most important capability for full-package firms, after the ability to source fabric, a function determined primarily by access to credit.

The sample room at DT was located inside the firm's main corporate building in its private trade zone on the outskirts of Santiago. Twenty-eight sewing technicians (*técnicos de costura*) worked in the department, making a dozen samples a day. The area was divided into two sides: the wovens section, where 12 male technicians sewed pants and shorts, and, across a small aisle, the knits section, where 10 male and 4 female sewers produced samples of casual tops. The work of the sample room sewers stood in marked contrast to the sewing that took place on the factory floor. It was slow and careful; every seam was sewn, then carefully examined, and often measured by a ruler. The sewers moved among all the operations in the labor of assembly, punctuated by the work of checking, correcting, and sometimes, to get it exactly right, hand-stitching. Margins between seam and hemline were measured over and over again; any more than a quarter-inch variance would spell rejection by the inspectors. Sewers here were paid a fixed salary with no production bonus. Carlos, an industrial engineer and the sample room supervisor, sat at a desk facing the two sample-making sub-areas. He complained that it was almost impossible to speed up the work without production bonuses. "They just don't have the attitude of professionals," he lamented.

Carlos' frustration pointed to the peculiar position of these skilled sewers who were neither assembly workers nor management. Many had painstakingly reversed the deskilling process of the progressive bundle system, learning all the operations needed to complete a garment. While their efforts had not yielded substantial changes in remuneration from the industrial system (the average weekly salary was 2,000 pesos), they had found ways to be less beholden to it. Nicolás was a good example. Forty years old and soft-spoken, he had started in the company in 1989 as a helper (*ayudante*), the lowest position in the industry, moving pieces between machines. "I dedicated myself to learning to sew," he told me, "using my lunch breaks to learn to use the machines." He eventually took a government-sponsored training course in cutting and sewing and, for the past several years, Nicolás had spent nights and weekends working as a custom tailor from his home. In addition to his steady salary at DT, he could make up to 1,500 pesos per week doing the additional custom work. A big problem with this strategy was electricity in his working-class neighborhood, where blackouts were frequent and often long lasting. To keep up with the custom work, Nicolás had invested in a generator and was paying it off monthly: 1,671 pesos every month. His wife also worked at DT but knew only one operation: attaching belt loops.

Like Nicolás, the handful of women in the sample room had trained themselves in trade zone factories over 10, 20, and even 30 years. Carla, for example, had worked for the firm for 14 years. She had started in inspection, a low-paying position occupied almost exclusively by women. When she had a child with her husband, a quality supervisor in the same plant, she requested an operator position because inspectors are among the last to leave the plant, and she found it difficult to leave late with the new demands that childcare placed on her time. Initially, her boss encouraged her to come in on Saturdays to learn other operations, but because of her childcare responsibilities, she explained, Carla declined. Instead, she used small breaks during the workday to learn to make the "J" operation on pants (i.e., the stitch in front of the fly) and then learned to sew pockets. Her husband provided the space and time for her to move to different machines and to learn additional operations until she could do enough of them to gain a position in the sample room. To supplement her factory salary, Carla sold creams, lotions, and hair products from a catalog to the managers and engineers in the corporate offices in the same building. On evenings and weekends, she took over childcare from her mother who looked after her three children during her paid working hours at DT.

The stories of Nicolás and Carla demonstrate that for sewers from the industrial system, movement between tasks was a key vehicle for learning multiple operations. Moreover, training depended primarily on informal relationships with supervisors to facilitate this process. In her ethnography of a Dominican export garment firm, anthropologist Portorreal (1991) observed the increased monitoring of women's movements on the factory floor, controlled by male line supervisors. Sociologist Leslie Salzinger (2003) also described how gender dynamics affected the mobility of men and women differently in a Mexican firm. The power relations between male line supervisors and the largely female production workforce circumscribed the mobility of women workers. The gendering of mobility within factories helps explain the absence of women in sewing positions that required knowledge of all operations, including sample room sewers and utilities. The perception of certain operations as more appropriate for men (i.e., attachment of waistbands) also played a role, since women who sought to train in woven garments would have to break into these male-dominated operations.

In addition to mobility within the plant, movement between factories was an alternative way for workers to train in different operations, learning a new operation at each new firm. Inter-plant movement was also gendered, reflecting men's and women's distinct labor market experiences. A survey of 1,500 garment workers in 2006 revealed that men were far more mobile between factories (i.e., were contracted a higher number of times) than were women. While 91 percent of men and 95 percent of women with five years of experience in trade zones were contracted one to four times, this number diverged as years of experience increased: at 18 years in the trade zones, 60 percent of men were contracted five or more times, compared to only 27 percent of women. In other words, the bulk of men with the same years of experience circulated through many more factories.[14] Nevertheless, movement between factories as a method for training was greatly inhibited by the employment crisis at the time of my research. Thus, given the trend of hiring workers with requisite skills and eliminating formal training programs for sewing, more men than women had skills in multiple operations because of the gender dynamics of both inter- and intra-firm mobility.

Inseparable from the informal, gendered channels through which workers transformed themselves into multifunctional operators or sewing technicians, perceptions of sewing skill also favored the notion that men were more appropriate for these positions. While sewers like Nicolás and Carla had worked their way up within the industrial system, nearly half the male sewers in the sample room were tailors. In the garment industry, the distinction between craft and industrial systems has never been hard and fast.[15] In the craft tradition, the gendering of sewing and skill extends from the garment itself: the labor of producing men's clothes, especially pants and shirts, is considered more skilled and thus also codes those who make men's clothes, tailors, as more skilled. Marcos, a sample room tailor, demonstrated this "fact" on a pair of US-brand men's pants by showing the various stitches that were exclusive to men's bottom garments. The work of tailors constructed sewing skill in contradistinction to the domestic labor of women's sewing. Julia, a middle manager, expressed this distinction in her reflections on sewing capacity in the Dominican Republic: "[H]ere [in the Dominican Republic], there is a sewing machine in almost all homes, at least there used to be because mothers used them to make little pajamas for the kids [and] grandmothers knit. In humble homes, women made sheets [and] patched clothes. There is a family tradition here but not an industrial one." The manager's comments revealed the broadly held belief that sewing was a naturalized skill, possessed by women who learn to sew at home. Collins (2003) has argued that despite the training that sewers must receive on industrial machines irrespective of their domestic "talents," this gendered presumption has long served to delegitimize women operators' skill-based claims in their demands for higher pay and collective representation, and, as I argue here, in their access to on-going opportunities for training in the workplace. With DT's efforts to forge a better position in the garment production network, the firm sought the acquired skills of men who were trained in the craft tradition, not the natural talents of women as domestic sewers (cf. Jenson 1989).

The narrative explaining the absence of women in more skilled sewing positions generally faulted female operators for not learning more operations. "They simply lack *chispa* [spark]," explained one sample room manager. Women like Carla were perceived to be exceptions. When I asked her why so few women were in the sample room, Carla reflected a similar perception to that of the sample room manager, explaining that women remained more dedicated to their homes than to paid work, a condition that "makes the country go backward." Both narratives reinforced the notion that skill was a matter of individual initiative and will, independent of gendered hierarchies of labor in the workplace and household.

Carla's narrative was reflected in the explanations given by industry spokespersons for the changing gender division of labor in the garment sector. For example, the director of the regional trade zone association discussed the barriers for women to become flexible operators in the following way: "The limits of women to train themselves [exist] for domestic and cultural reasons. This includes the attitude that 'I operate this machine and I'm satisfied with that because it's what I do. I don't want to sacrifice time with my kids' and all these cultural questions that impede that these multifunctional operators are women."

The explanation of women's absence from the upper rungs of the new skill hierarchy as "cultural" in effect ascribed to women the quality of rigidity, a quality that was characterized by a refusal to govern the self in accordance with the requirements of the new production model. Distinct from the younger and largely female workforce idealized as flexible under the strict assembly model, women workers in this context were now perceived as unwilling to make their own labor flexible. As I have demonstrated, this perception masked the gendered politics of mobility within and between firms, the gendering of household labor, and the association of masculinity with skilled sewing through the craft tradition. Enmeshed with these barriers to self-training, we should not discount women workers' possible resistance to the new production regime reflected in the narrative of refusal offered by the industry spokeswoman. The outcome of these interwoven factors was to legitimate the changing face of the garment workforce in the upgraded firm as more men were incorporated into skilled sewing positions and as women were increasingly excluded or relegated to lower-paid jobs.

Let me now turn to the second example: the feminization of engineering. The case of engineering stands as an important counter example, illustrative of what some scholars have observed as the diverging fortunes of middle- and working-class women in the Caribbean (e.g., Mullings 2005). In the Dominican Republic, women make up 65 percent of college enrolment. Although women constitute only 24 percent of students enrolled for engineering degrees, industrial engineering is reported to be "highly feminized" (Quiroga 2003; see also Hualde Alfaro 2001). In my observations at DT and two other full-package firms, I found that as firms repositioned themselves in the garment production network, industrial engineering became associated with new responsibilities, and the workers who were perceived to be best suited to undertake these roles were women.

The principal role of the industrial engineer in assembly production is to arrange sewing operations by calculating and distributing standard allowable minutes on the assembly line. As I noted in the previous section, engineers at DT continued to organize assembly time but allocated more operations to each operator. With the shift to full-package supply, however, more industrial engineers were hired to act as links between the firm's foreign, primarily US buyers and the production process. The responsibilities of these engineers varied across firms, depending upon the degree of vertical integration of the operation. At DT, these engineers were responsible for sourcing materials, managing those materials through the production process, and adjusting orders during production to correct errors or to accommodate mid-order changes requested by the buyer.

The shift to full package at DT saw the growth of so-called front-end and back-end support teams, a new structure at the firm staffed by these engineers. With one exception, all these engineers were women. Such an observation is insufficient, of course, to understand the gendering of this function and what I observed as a feminization of engineering at the firm. One can begin to understand the gendering of this new role through the narrative of one of these engineers. Veronica, a woman in her early thirties, was the head of the support teams for the wovens division. Citing the example of a male co-worker who had started at the firm in the sourcing department and left once the team structure was in place, Veronica explained why, in her opinion, women filled these positions: "I can tell you that 99 percent, no 100 percent, of the people who are doing planning in [the firm] are women because [the work] requires a lot of details ... this position requires a level of details and of handling, coordination at a detailed level, that men have many good qualities and better [ones] than women, but women have a better handle on details." Skill in this instance was naturalized through the association of women with "detail work." Moreover, the shift to the team structure reordered functions that were previously managed as separate departments – such as sourcing fabric and trims, planning, and client relations – all under the rubric of client services. It was not that the work was necessarily more or less detailed now that it was organized into teams; rather, this reorganization of functions also resignified this work from production to services. The work appeared subsequently to be reproduced as feminine and filled by female engineers.

Elements of class difference underlay this changing role of engineers in the upgraded firm. None of the service engineers whom I spent time with in DT and in two other firms had much direct production experience, for example. They entered the industrial system through college-related internships. In contrast, the male production engineers whom they supervised often had significant production experience: several of those who I interviewed had started as operators while they completed their technical degrees. Knowledge of English was another marker of class distinction between the service engineers, who were all bilingual, and the production engineers, who were not.

Tension between the service and production engineers was easily identifiable. In one meeting, in which a service engineer, Stacey, was questioning a plant engineer, Rodrigo, about the percentage of thread and cloth waste in the plant, Rodrigo asked Stacey if she had ever worked on the factory floor. Stacey responded that she had, for three years (this was part of her internship program), and Rodrigo responded that he had done so for 14. Stacey needed to get on the machines to understand how much thread they used, he concluded defensively.

Managing production was something that the service engineers admitted to being nervous about as women in supervisory positions. For example, Veronica recounted her experience of being hired in the following way: "When they told me [about the job], I said to them, are you sure you want a woman in this position? I never thought that they would give this job to a woman. I have men under me, and a production plant because the cutting plant reports to me ... I never had the opportunity to run a plant at the operations level before. I like it. My boss says that I was born a woman, but I think like a man. I tell him no. I am a woman even if he thinks that I think like a man." Veronica's observations suggest the desired qualities of these new engineers on the part of her superiors in upper management at DT, who remained entirely male (except for a female family member of the owner). Feminized in their attention to detail and their ability to attend to buyers' requests, service engineers were encouraged to "be manly" when dealing with their male subordinates in the production plants.

A feminist analysis reveals the importance of gender norms in the reworking of "skilled" and "unskilled" work in the labor process of full-package firms. As Jane Collins writes, skill is "a dense nexus of claim making" (2003: 171), wherein workers assert the scarcity and value of their work as skilled, while managers attempt to shape and undermine skill claims to create occupational hierarchies. The naturalization of sewing as an ability found among women served to define the predominantly female assembly workforce who sewed in the trade zones as unskilled prior to restructuring. Sewing was an ability that was abundant, required little training, and received low remuneration. With the restructuring of full-package firms, management implemented a new skill hierarchy through a reworking of gendered understandings of skilled and unskilled sewing. The cultural construction of female workers as resistant to the new paradigm of "multifunctionality," combined with the more limited mobility of women on and between shop floors and received biases in the craft tradition, resignified the labor of sewing in restructured garment firms as a masculine activity. This process of gendering created some opportunities for men – albeit limited and not well paid – to scale the new sewing skill hierarchy, while gradually excluding women garment workers who performed a single operation.

In short, Santiago garment owners' efforts to distinguish themselves from their foreign competitors – whom they characterized as majority female-employing, footloose, and low-tech – were based fundamentally on the use of gendered norms to rework occupational hierarchies. The result was the production of new subjects in restructured garment firms: the masculine, multifunctional "sewing technician" and the feminized, college-educated "services" engineer. This process of internal restructuring was inseparable from new relations of subcontracting and outsourcing, such as the relocation of assembly functions to arm's-length contractors regionally, and, as we will see in Chapter 5, to Haiti. In short, garment restructuring was not only about firms scaling the value hierarchy of the global production network. The latter depended upon management's ability to reorganize the labor process, thereby reworking the internal occupational hierarchies of the firm. As we have seen, gendered practices and norms provided reason and meaning to the reworking of labor's value, while simultaneously making this process appear to be natural. These observations offer an important, sympathetic corrective to recent feminist studies of subject production in global factories (e.g., Salzinger 2003). We must observe not only how certain kinds of workers are produced (or not) on given shop floors (as disposable, as feminized, etc.), but also how these very categories are reproduced through restructuring processes linking shop floors in global production networks.

## Gendered Economic Dualisms: Enabling Export Restructuring as Development

While the previous two sections focused on the discourses and practices of restructuring in the Dominican garment sector, in this section I discuss the narratives of international development agencies, specifically the World Bank. By the 2000s, with increased global competition and stagnant manufacturing export growth in the country, World Bank experts sought to redefine the export-led development model long promoted by Washington-based policymakers. The premise of my discussion of these World Bank reports is that development discourses do not merely describe the "Dominican economy." These discourses are fundamental to shaping the policies and metrics that condition flows of investment and influence the politics of value that sustain and reproduce the export paradigm.

By reading development texts next to the narratives we have just
encountered in the garment sector, we find an apparent dispute between
the industry and policymakers. On the one hand, the World Bank's
analysis of the sector in the mid-2000s appeared to be at odds with the
claims to the high-tech t-shirt asserted by Santiago garment owners. Just
like the Dominican government's competitiveness report and its assertion
of a transition to *mentefactura,* World Bank experts evaluating the
Dominican export portfolio also concluded that the garment sector – and
the institutional structure of trade zones that supported the industry –
was part of the Dominican economy's past. Policymakers' ability to
construct a naturalized, sequential transition beyond trade zones relied
upon gendered understandings of labor's value similar to those deployed
by the industry. The World Bank's claims mobilized the long-standing
development construct of a "dual economy," a hierarchical, binary model
that defines and fixes relative categories of traditional and modern, and
of value and non-value, in order to produce stage-like development
sequences. Gendered notions of labor's value/non-value were hinged to
this binary dualism of economic value/non-value. Thus, if, in the case of
the Santiago garment industry, gendered discourses and practices served
as a resource to flexibilize labor in novel ways, we will see how, in the
case of the World Bank, a gendered dual economy lent flexibility to
neoliberal development policy. In both instances, whether in defense of
the industry or the opposite, these discourses served to naturalize the
expulsion of feminized labor.

Writing in the wake of the lost decade of the 1980s, the World Bank
characterized tourism and trade zones as the two sectors that had
withstood the vicissitudes of the tumultuous period of stagnant and
shrinking economic activity following the macro-region's debt crisis:

> The Dominican Republic has a strongly dualistic economy – a highly pro-
> tected domestic sector which has deteriorated sharply since the early
> 1980s and a non-traditional, export-oriented, private-sector-led sector
> [tourism and FTZs] which has continued to perform well even during
> adverse periods. The contribution [of FTZs and tourism] attests to the
> high potential of the Dominican economy to compete in external markets.
> (1995: 1)

The quote and subsequent analysis in this report and elsewhere set up
the notion that the country was comprised of parallel, opposed economic
spaces and paradigms: a domestic economy, burdened by the public
sector and protected ISI industries, and a dynamic "private-sector-led"
or market-led sector. This construct recuperated the notion of a dualistic
economy, which has long served as a powerful discourse to frame
market-oriented policy interventions.

Dual economy discourses draw upon the Western epistemological structure of binary dualisms to resolve a fundamental dilemma for the proponents of capitalism as a universal model of development: if the market is natural, and thus should expand on its own, why does it fail to do so? In short, why do we need "development"? St Lucian-born and British-trained economist W. Arthur Lewis, a key architect of "industrialization by invitation" and advocate of the model in Puerto Rico (see Chapter 2), introduced the dual sector model to address this conundrum. Lewis argued that "developing" economies consisted of a modern, capitalist sector driven by competitive forces and generative of surplus and thus accumulation, on the one hand, and a subsistence sector, "unproductive" by definition because the sector "is not fructified by capital," on the other (1983 [1954]: 147). Lewis famously wrote that "[e]ven in a very highly developed economy, the tendency for capital to flow evenly through the economy is very weak" (1983 [1954]: 148). "In a backward economy," he continued, "it hardly exists. Inevitably what one gets are heavily developed patches of the economy, surrounded by economic darkness" (Lewis 1983 [1954]: 148). Lewis advanced the notion that the capitalist economy required interventions to "conquer" the darkness, while concretizing the idea that these two spheres represented a past and a future along a continuous trajectory of Progress (Escobar 1995). Economic dualism justified the need to plan capitalist expansion, while recuperating the idea of capitalism, and the competitive, surplus-producing market relations that define it, as both a natural state and telos of development.

The binary structure of the dual economy repeats – with difference – foundational tropes of the colonial/modern project of the Enlightenment. These include the hierarchical, binary structures of masculine/feminine, modern/traditional, and mind/body, which share not only the same structure as, but also serve to sustain, the binary opposition between "modern" markets and economic "backwardness" (see Gibson-Graham 1996). Timothy Mitchell (2007) has argued that the key function of dual-sector discourses is to construct a boundary to the market, which subsequently becomes an object of development. "For countries outside the West," he writes, "the idea of a boundary provides a common way not just to think about these places but to diagnose their problems and design appropriate remedies" (Mitchell 2007: 246–247). The construction of a limit between the market and non-market, conflated with capitalist and non-capitalist spheres, has long served to rationalize interventions that are thought through the notion of a transition from non-market/non-capitalist ways of organizing production, provision, and returns to market/capitalist mechanisms. This transition has never been solely temporal, however; the construct of a transition to capitalist markets has always also contemplated new spatial arrangements. In the mid-twentieth century, for example, this transition was conceptualized through the conflation of development with urbanization.

Of course, the "dual economy" posited by the Bank in the Dominican Republic was patently *not* equivalent to the capitalist/non-capitalist model proposed by Lewis and critiqued by Mitchell as the market/non-market horizon. Rather, the Bank's discourse signaled a dual structure of market *regulation* associated with the competing paradigms of Latin American capitalist development: import substitution (the "domestic sector") and export-orientation (i.e., trade zones and tourism). Nonetheless, the idea invoked an analogous hierarchical, binary structure to powerful effect, as we will see. Like the dual sector model, the Bank's construction presented these two "economies" as distinctly oriented and minimized the mutual connections between the two, while simultaneously constructing a strict state/market divide corresponding with the two sectors. The so-called domestic, state-led sector was the unproductive, moribund, traditional other to the modern, market-led export sector. This conception erased the fact that the "domestic sector" had long been deeply dependent upon the exports of sugar and ferronickel, with varying degrees of state and private capital involved. Moreover, the "private sector-led" sector, dominated by garment exports, was the outcome of a highly regulated form of international trade managed by states through quotas, as explained in Chapter 2. Finally, the most dynamic trade zone region, the Cibao, drew upon long-standing resources, skills, and class relations inseparable from the region's role in so-called traditional agro-commodity exports – tobacco, cacao, and coffee – and the region's subsequent incorporation into state-subsidized import substituting agro-industries (see Chapters 2 and 4).

The elision of these connections between the country's supposed "two sectors" and the complex state–market relations that were central to both forms of regulation facilitated the main *effect* of the Bank's dual economy discourse: the notion that the enclave sectors were measures of the Dominican economy's *potential*, a potential constrained by the "low productivity and anemic growth" of its domestic part (World Bank 1995: ii). This policy discourse succeeded in positioning trade zones, along with tourism, as the positive polarity of a binary dualism and as a symbol of modernization and progress. The authors argued that in order to realize its potential, the so-called domestic sector would have to "emulate the two enclave sectors [i.e., tourism and trade zones]" (World Bank 1995: ii). "To achieve this goal," they concluded, "[the country would] require a stable macro framework coupled with a deepening of recent structural reforms" (World Bank 1995: 3).[16] These policy prescriptions included privatization of public utilities, liberalization of investment, tariff reductions, and downsizing of the state bureaucracy. The proposed reforms were the same hard-core nostrums associated with the Washington Consensus, or what Peck and Tickell (2002) have usefully dubbed as "roll back" neoliberalism, or a first phase of market-led regulation that sought to position markets as determinants and arbiters of policy (see also Chapter 2).

The reliance of the new export sectors on the structurally devalued labor of women workers was not mentioned explicitly in World Bank analyses of the Dominican Republic at the time. The fact that both tourism and trade zones provided employment, however, was celebrated as a positive outcome that gave further credence to the model:

> The performance of tourism and FTZs – two labour intensive, export-oriented sectors – indicates the potential impact of outward-oriented growth on employment and poverty. Since wages are the main source of income for the poor, the link between poverty and labor-intensive growth is clear and direct. (World Bank 1995: 28)

Again, the "market-led" sectors were windows onto a future Dominican economy where high growth stimulated the expansion of employment for the poor, resulting in the reduction of poverty.[17]

A decade later, the dual economy discourse persisted and the accounting techniques to inscribe the dualism were even more robust. By 2006, the dualism was conceptualized as two distinct concept-spaces: the FTZ economy and the "non-FTZ economy" (see Figure 3.2). The differences between the two were made evident through constructing isolated and

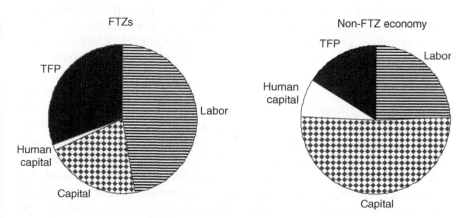

**Figure 3.2** The Dominican Republic as a dual economy. Source: World Bank Country Economic Memo, 2006: vi. Reprinted by permission of the World Bank.

comparable measures of their component parts: productivity, labor, human capital, and capital. By showing differences between these measurements – for example, the predominance of "unskilled" labor in trade zones versus so-called human capital in the "non-trade zone economy" – the Bank asserted the nonintegrated and bounded nature of FTZs and "the rest of the economy."

The Bank's findings with respect to the dualism initially appeared to differ little from the decade before. The authors argued that FTZs benefited "from the dynamism stemming from competition and integration with external markets ... [While] the domestic economy, by contrast, [followed] a traditional model of capital accumulation and import substitution, and thus [was] unable to compete globally" (World Bank 2006: vi). Yet, despite this familiar story, the Bank drew markedly different conclusions from those drawn the previous decade. In fact, the enclaves were no longer *windows* onto market potential but rather *burdens* to "the rest of the economy." The Bank asserted:

> [T]he strategy of FTZ-based export promotion and the all-inclusive enclave tourism model effectively inhibited more efficient domestic production and the positive interaction between export and domestic production, thereby reducing the potential growth pay-off from trade openness. (World Bank 1995: x)

The shift in the Bank's evaluation was consistent with new development policies that were moving away from the "roll back" approach of the Washington Consensus associated with freeing markets from the burdens of the state, and toward an emphasis upon creating positive conditions for the functioning of markets through a pliable notion of market governance (Craig & Porter 2006). In this context, free trade zones as spaces of market dynamism were resignified as burdens upon market development within ("the rest of") the Dominican economy. The economic unevenness diagnosed through the Bank's dual economy accounting now indexed the poor functioning of trade zones to distribute the gains of the "market-led" sector throughout the economy. The problematic, framed in this way, repeated the structure of Lewis' model of the 1950s, albeit in a context where markets predominate. The "market" sector had failed to simply spread and thus transform the "backward" sector through first wave structural adjustment reforms. Instead, the authors argued, only through targeted interventions would the competitive market burst the banks of the enclaves to transform the lagging "rest of the economy."

How did the resignification of trade zones from windows of market potential to burdens upon market development occur? The dual economy discourse constructed by the Bank in the mid-1990s rested upon the notion that FTZs were circumscribed spaces where the fetters upon the free market were loosened, allowing for the latter's natural development. In this way, trade zones were portrayed as robust contributors to an economy otherwise in crisis: the authors characterized the economic performance of the country in the 1970s and 1980s as poor, with average growth of 2 percent a year and an overall decline in GDP per capita over the same period. The difference in economic performance between FTZs and the "domestic sector" thus served to demonstrate constraints upon the latter, attributed to state interventions that "weighed down" the natural market lying smothered beneath.

By the mid-2000s, however, rather than jump-starting a market economy or indexing its (lost) potential, the Bank represented trade zones as hindering the realization of market liberalism in the Dominican Republic. To put it succinctly, trade zones had themselves become fetters upon the natural development of markets. The Bank now characterized the Dominican Republic as having "strong growth outcomes," well above the rest of Latin America and the Caribbean. The high growth of the 1990s was projected back onto the previous decades. The Bank thus described the country as benefiting from "[r]obust long-run growth tempered by several short-lived crises" since the 1970s (World Bank 2006: 13). While foreign investment in trade zones had demonstrated the potential of markets a decade earlier, in the high-growth narrative of the mid-2000s, the Bank argued that the channeling of foreign investment into trade zones was a constraint on the functioning of markets. Recast in the new story of robust and sustained economy-wide economic growth, the Bank concluded that trade zones had failed to contribute substantially to the country's economic success. In fact, FTZs, the Bank asserted, were "likely to have constrained growth potential" (World Bank 2006: 20). This failure was not due to the small size of FTZs, which would have been closer to the argument for emulation of the 1990s, but rather to the existence of FTZs themselves, which were a hindrance to market formation through their distorting incentives which channeled investment towards them.

An important justification for this shift rested precisely upon the idea that the mass employment of women in low-wage, so-called unskilled sectors was a drain on market-led development. In contrast to the argument a decade prior that employment of the poor constituted a win-win for market-led regulation and welfare, the Bank now argued as follows:

> [E]ven though FTZs created thousands of jobs, employing lots of women in low-skilled work *who might have otherwise been out of the labour force* or engaged in even lower-paying activities, the spillovers to the rest

of the economy would have been greater under higher investment in technology and skills to support higher value-added production. (World Bank 2006: viii–ix, emphasis added)

The former benefits of employment in trade zones as portrayed by the Bank were reworked as evidence of a kind of market distortion trapping capital in spaces of "low-value," made possible through the distinction between "labor," on the one hand, and "human capital" on the other. The former category was signified by female waged workers whose wage employment, the Bank implied, was more of a temporary aberration. In this discourse, the incorporation of women workers into trade zones was a kind of exception to the fundamental construct of female labor as non-value, as un-employed (Federici 2004). This basic structure of coloniality, as I discussed in the introductory chapter, could be recuperated here as a resource for justifying *ex post facto* the limits of roll back neoliberal strategies of the 1980s and 1990s. In resignifying trade zones from market potential to market burden through economic dualism and female labor's non-value, the market could be recuperated as a foundation for development, even as the shortcomings of earlier neoliberal policies were critiqued by their erstwhile proponents. In short, these binary, gendered constructs of labor's non/value and economic non/value lent flexibility to regulatory restructuring, reproducing, rather than challenging, the neoliberal paradigm.

There is a certain irony to this discursive construct: although the majority of workers expelled from the trade zones during the mid-2000s employment crisis were women, a high proportion of low-wage men also lost their jobs. What is clear from this resignification of trade zones from market potential to market burden is that the imperative of sequential development to *mentefactura* rested upon a feminized body of labor whose disposability was constructed as both necessary and natural to secure the country's progress (cf. Wright 2006).

## Recuperating the Gender Politics of Uneven Development

As I argued in the introductory chapter, the disintegration of production into functions that are carried out by a transnational network of suppliers reproduces hierarchies of value, defined by mixes of periphery and core functions. In Santiago, we saw two rungs of this hierarchy: the highly competitive, low-margin position of assembly-only suppliers, on the one hand, and the semi-monopolized position filled by full-package firms, on the other. The efforts of firms like DT to access the middle and upper echelons of this hierarchy required the redefinition and the relocation of "skilled" and "unskilled" work, and the exploitable subjects of these new arrangements. These transformations depended upon gendered discourses and practices in several ways. In Santiago, as a select number of suppliers sought to establish themselves within higher value niches of the garment production network, managers and others drew upon gendered norms of labor's value to reshape the workplace occupational hierarchy based upon new definitions of skill. Gendered norms served to rationalize the resulting shift in the dialectic of embodied labor and abstract value, which materialized in the incorporation of large numbers of men into surviving garment firms and the expulsion of feminized labor. In short, production network restructuring took place through a reworking of the techno-organizational division of labor – which firms do what and where – *and* the dialectic of embodied labor/ abstract value. As Elson long ago argued, labor and labor process functions are mutually constituted as labor *becomes* socially fixed for a time (1979: 124–130; see also note 3).

In short, the global factory materializes a temporary social and spatial arrangement of functional cores and peripheries and exploitable workers. Gender norms serve not only as a resource to rationalize firm restructuring, but they also lend meaning to these changes expressed in stage-like terms, as natural progress of "the market." I have made this latter case by revisiting the critique of dual economy models in the context of regulatory flexibility, or the reworking of market-led neoliberalism in the face of neoliberal policy failures. Development experts were able to sustain the notion of the market as natural, and the end-point of a teleological process of capitalist development, even as export-oriented enclaves reached their limits of profitability and failed to act as catalysts of an idealized market competitiveness. As we saw, the shift in signification of trade zones from potential to burden was inseparable from a resignification of labor's value by naturalizing the expulsion of feminized labor. Fundamentally, the construction of trade zones as regulatory burdens had tangible effects; the Bank's objective was to erode the system of incentives and preferences that the garment sector was lobbying to maintain. Despite this apparent conflict, both garment owners and development policymakers drew upon gendered constructs and practices to make their value claims: the exclusion of women workers from the "value-added" companies in Santiago, on the one hand, and the symbolic expulsion of feminized labor in the World Bank's development narrative, on the other.

In this chapter, I have largely told the story of sector restructuring, the gendered politics of labor's value and the dynamics of uneven development from the perspective of capital. If capital's imperative to renew accumulation through the restructuring of production networks drives the on-going process of making uneven development, however, it is only in part. Dominicans and Haitians have created their livelihood possibilities through place-specific strategies of accommodation and resistance. These strategies have reflected and reshaped the two countries' complex and relational trajectories of change and their particular subnational regional formations. In the next two chapters, then, we turn to the experiences of unemployed garment workers in the Cibao, and workers and community activists at the Haitian–Dominican border. Their stories and strategies shed light on how individual and collective efforts to adapt to, and to contest, the hierarchies of value that sustain global production forge geographies of uneven development.

## Notes

1   Encuentro con legisladores, Santiago textile training center, Santiago FTZ. October 30, 2006.
2   This chapter contains some materials previously published in Werner (2012).
3   In her famous intervention into Marxist debates on the labor theory of value, Diane Elson proposed a "value theory of labor." In essence, she argued that "value" – that is, labor – is not the origin or cause of prices and profits. Marx's intervention was fundamentally to shift the object of political economic analysis from exchange to production; thus, the object of his analysis was to understand why and how labor *becomes* value. In short, Elson writes: "It is not a matter of seeking an explanation of why prices are what they are and finding it in labor. But rather of seeking an understanding of why labor takes the forms it does, and what the political consequences are" (1979: 123).
4   All four of the product development departments where I observed operations had hired New York City-connected veteran pattern-makers not only to train their own pattern-makers, but also to lend confidence to their clients, the US brands and retailers. Most of the US-based pattern-makers stayed for three years and left. In one firm, the pattern-maker stayed on for another four years and coincided with my research. Harold, a Colombian-born man in his mid-fifties, the son of a medium-sized garment factory owner, had spent three decades working in the industry between New York City and Miami, including work in the sample room for Liz Claiborne. In the Dominican Republic, Harold spent part of his days in the product development department of one of the major full-package, Santiago-based companies. While he had long finished formally training their staff, Harold told me that the owner had kept him on because Harold could tap into his

New York network, including long-time associates still working for Levi's, to troubleshoot problems.

5  These figures are calculated based on unpublished employment numbers facilitated by the three firms and government figures for total garment employment published by Consejo nacional de zonas francas de exportación (CNZFE; various years).

6  This regulatory change was the passage of so-called NAFTA parity, formally called the Caribbean Basin Trade Partnership Act. See my discussion in Chapter 2 for more on US trade policy and its limitations.

7  Informants consistently estimated the cost of cloth to be around 60 percent of the cost of the garment. The difference in operating capital between an assembly firm and a full-package firm is more substantial. One owner of a medium-sized firm that did both contract assembly and full-package work gave the following estimates: to assemble 11,000 dozen t-shirts, the firm required 180,000 dollars of operating capital; to produce 9,000 dozen as a full-package operation, the firm required 2.5 million dollars.

8  A survey of 1,500 garment workers in 2006 confirmed significant differences in pay among operators in the export garment sector: 35 percent of women compared to 17.5 percent of men earned 3,600 pesos per month or less (USAID 2007: 35). In 2006, 1 dollar was equal to 28 pesos.

9  Moreover, considering that trade zone workers worked an average of 5 percent more hours per week (World Bank 2006: 92), wage convergence should not be overstated.

10 On the gendering of work in cutting rooms and the role of automation in Mexican *maquilas*, see Fernández-Kelly (1983: chap. 1).

11 For example, an order of 3,000 pieces per week could be filled by five daily quotas of 600 pieces by combining operations, rather than three daily quotas of 1,000 pieces performed in single operations followed by a change of style or a layoff for the remaining two days of the week.

12 If a Level 1 operation was combined with a Level 4 operation, for example, the worker earned the rate corresponding to the higher level (i.e., Level 4).

13 Workers confirmed this change in company practice. I discussed the company's history of training at length with its long-time head of training who confirmed the phase-out of company-sponsored sewing training in Santiago. In line with this change, he was transferred to the firm's new operations on the Haitian–Dominican border in 2003.

14 Calculated from the National Survey of Displaced Trade Zone Workers (on file with author). See USAID (2007) for additional information.

15 See Tewari (2008) on tailors in Indian garment export firms; also Narotzky and Smith (2006) on the amalgam of craft, domestic, and industrial practices that comprised the Valencian garment sector.

16 The authors placed great emphasis on emulation and argued against policies that would integrate trade zones and tourism through linkages to domestic producers. The rationale for their argument (emulation over integration)

was the small size of the enclaves; the logic, however, underpinned their argument for the core policy prescriptions of the "roll back" phase of neoliberalism.

17    The mention of poverty in this "roll back" neoliberal example should not be confused with the subsequent mainstreaming of poverty reduction at the World Bank as part of the institution's efforts to address the critiques of the Washington Consensus. In this case, the authors presumed poverty would be reduced by a basic market mechanism – that is, the expansion of employment to the poor. In the so-called post-Washington Consensus period, the World Bank purported to address why this and other market mechanisms had failed to reduce poverty and inequality. In general, the Bank's new focus on poverty reduction did not replace its emphasis on markets as a goal of development, but rather sought to complement this goal by foregrounding social concerns that would enhance or improve the outcomes of market participation (Bergeron 2003; Wade 2002). For more, see my discussion of the international financial institutions' Poverty Reduction Strategy Paper in Haiti (Chapter 6).

# 4

# Embodied Negotiations
## Geographies of Work after Trade Zones

At the end of 2006, one of the three largest Santiago-based garment firms, IA Manufacturing, announced it would be shuttering its operations, idling a considerable portion of the Santiago trade zone where eight of its plants were located. The former chief financial officer attributed the closure to a severe crisis of liquidity brought about by the company's inexperience in managing textile inventory, fluctuations in currency value, and difficulties meeting the price demands made by the firm's main client, Levi's. Between the shutdown of the IA Manufacturing plants, an additional 15 other factory closures, and the suspension of operations at another four, the zone had a spectral feel. Lone security guards sat outside vacant buildings, many housing machines embargoed by creditors or in the process of sale. Old presses draped in shredded blue tarp sat on silent loading docks and dust-caked lunch tables stood empty next to dented lockers. On Monday and Tuesday mornings, groups of job seekers made almost ritual rounds in the zone to look for work at fewer and fewer factories. The sounds of steam trouser presses and nostalgic *bachata* music that emanated from the few operating plants were quickly subsumed by the idled space surrounding them.

In this chapter, I tell the story of global production from the threshold of the idled factory.[1] I draw upon an instance of factory flight – the closure of IA Manufacturing – to refocus our understanding of global production upon the navigation of the boundary between wage labor and non-wage work. If the exploitation of labor power is the basis of capitalist production, the reproduction of capitalism depends upon articulating the wage relation with a host of other arrangements of work

*Global Displacements: The Making of Uneven Development in the Caribbean*, First Edition.
Marion Werner.
© 2016 John Wiley & Sons, Ltd. Published 2016 by John Wiley & Sons, Ltd.

(Quijano 2000a; 2000b). Feminist scholars have long explored how women's unpaid domestic work is central to reproducing workers as labor power during moments of capitalist expansion. This intimate nexus of wage labor with other work arrangements also makes the process of capitalist restructuring possible. The devaluation of labor at the heart of capitalist restructuring fundamentally relies upon the complex livelihood strategies that the working poor craft in order to reproduce themselves through tumultuous rounds of accumulation and crisis.

In this chapter, then, I explore these livelihood strategies through the narratives and practices of former trade zone workers, or *zoneros*, as they are commonly referred to in the Dominican Republic. To do this, I draw upon Judith Butler's notion of performative subjectivity. Butler (1993) argues that the body materializes as a subject through processes of boundary-making that simultaneously mark an apparently purified subject field, the normative subject, and the subject's other, the repudiated or the abject, expulsed subject. As Melissa Wright (2006) explains, all bodies are composites of multiple identities, but to be produced as the source of surplus value, these bodies must materialize as the transparent, normative subject of value – that is, as labor power. This process of subject production always also produces its other: waste, disposability, the *un*employed – that is, working bodies literally assigned the status of "non-value" within the logics of capitalist production.

Unemployed workers experienced the dialectic of value and waste that propels capitalist accumulation as a set of struggles and strategies to reproduce themselves as subjects of social worth. While wage work remained an often-unattainable ideal, especially for women as we will see, former *zoneros* navigated uneven geographies of livelihood to forge social positions that embodied cultural notions of worth. These positions of social worth were inextricably bound to ideas of modernity and progress, even as workers faced temporary or permanent exclusion from the wage relation. I argue that such negotiation of social position by embodied subjects *is always also* a process of navigating and forging a spatial position in localities where labor is simultaneously raced and gendered. Here, then, I bring together Butler's and Wright's insights on subject production with a postcolonial notion of geographies of work. Geographies of work aim to understand the broader field of "innumerable and meaningful acts of fabrication that make social life possible" (Gidwani & Chari 2004). In doing so, they expand the field of our understanding of working subjectivities from the classic factory ethnography pioneered by Michael Burawoy (1979) – who famously asked why industrial workers consent to work as hard as they do – to the subjectivities that are produced through navigating the very instabilities of capitalist production.[2] Such an approach is profoundly commensurate with Caribbean and post-emancipation societies more broadly, where subaltern populations have long expressed deeply contradictory relationships with, and oftentimes outright resistance to, proletarianization (Cooper et al. 2000).

In recounting former garment workers' experiences, I focus on the struggles of Dominican migrant workers who navigate livelihood options between their rural hometowns, or *campos*, and the city of Santiago where many had worked for well over a decade or more. I term their on-going efforts to forge positions of social worth *embodied negotiations*. Workers' efforts to create social and spatial distance from abject "others" are simultaneously expressed in claims to occupy modern social positions and to embody cultural notions of progress. These embodied negotiations are conditioned by, and simultaneously seek to redefine, the gendered and racialized hierarchies of labor that pervade the urban and rural contexts of the Cibao. In short, they constitute a form of subject production that does not simply occur in space but rather is dynamically articulated with, and is part of remaking, sociospatial divisions of labor. In short, the restructuring of production networks is not solely the product of the simultaneous reworking of firm functions and the production of subjects for these occupations, as we saw in the previous chapter. The devaluation of labor at the heart of global production restructuring brings to the fore the wage/non-wage boundary. Embodied negotiations of this boundary by the unemployed simultaneously rework racialized and gendered regions, cities and households, thus reproducing uneven development anew.

In order to develop this argument, in the following section, I describe the changes in regional accumulation strategies and migration flows that transformed the Cibaeño *campo* and the region's commercial capital, Santiago. I focus on the period of adjustment to so-called non-traditional exports, especially export garment factories, beginning in the 1980s, and analyze dominant spatial discourses and practices surrounding the sector's employment crisis in the mid-2000s. The purpose here is for the reader to understand how raced and gendered hierarchies of labor have been reworked spatially as the region has been transformed by repeated waves of accumulation, dispossession, and devaluation. In the third section, "Workers' Strategies as Embodied Negotiations," I present the strategies of former garment workers to constitute livelihoods by navigating these racialized and gendered geographies of the region and the city. While, as I have argued, coloniality provides the basic structural underpinnings of the region's race, class, and gender hierarchies, my account offers an explanation for how these differences are contingently reworked. Workers not only struggled to make a living, but also to embody culturally constructed meanings of modernity and progress – that is, to forge positions of social worth. In the conclusion, I argue that these geographies of work can shed light on alternative forms of subject-making germane to the racialized and gendered hinterlands of wage work. In short, feminist scholarship on global production can be mobilized to account for how subjects are produced as much by the wage relation as by the social practices to navigate its very instability.

## The *Campo* and the Making of an Immigrant City

The *campo* is a polysemic term with more than 18 meanings in the *Real Academia Española*. In its use in the Dominican Republic, the *campo* can refer to rural areas or the countryside (*el campo*), a person's natal home in a rural area (*mi/su campo*), and a cultivated field (*un campo*). The last two meanings indicate the inseparability of family dwellings and agricultural production. Yesenia, a former trade zone worker, described the meaning of the *campo* to me primarily in terms of the location of her family: "The *campo*, the maternal house, the house where one was born. Your *campo* of origin, where your family (is), your father and your mother." Other workers used the term to distinguish themselves from co-workers whom they claimed were from places that were "more *campo*" than their own places of origin, referring to the perception of rural places as isolated or lacking certain markers of progress such as a gas station or a restaurant.

The rich symbolic valence of the term *campo* is an indicator of the complex trajectories of social change that shape the Dominican countryside. As I discussed in Chapter 2, the Cibao region has long been the paradigmatic locus of both the *campo* and the ideal subject of the Dominican nation, the surplus-producing Dominican male peasant. During the 31-year dictatorship of Rafael Leonidas Trujillo, about half the population of the Dominican Republic lived in the Cibao. Much of this rural population was settled on *minifundios,* or small landholdings, which Trujillo's government distributed in one of the largest projects of land reform in Latin America and the Caribbean (Turits 2003). The last decade of the Trujillato was witness to a shift away from small-farmer production, however, as the dictator sought to intensify agricultural output in order to supply food to the capital city, where he was promoting domestic import substituting industries, and to the sugar plantations of the southeast, which he had nationalized and greatly expanded. The gradual move away from state-support for smallholder production only intensified after Trujillo's assassination in 1961. Detrimental market terms for the region's traditional export crops – tobacco, coffee, and cacao – increased the number of sharecroppers, while the state's withdrawal of subsidies and other forms of support to rural producers further hastened the process of rural dispossession (Dore y Cabral 1981; Betances 1995; San Miguel 1997; Lozano 2001).

A substantial boost to the urban minimum wage and a loosening of state control over internal migration during the post-dictatorship transition period led to mass out-migration from Dominican *campos* by the end of the 1960s (Lozano 2001). Migration from the Cibao appeared to have a stepwise form: land- and capital-poor rural migrants from the surrounding agricultural region migrated to Santiago, while migrants from Santiago and the larger commercial towns of the central Cibao moved to Santo Domingo (Ramírez et al. 1988). All told, the Cibao experienced net migration on the order of a quarter million people annually through the late 1970s and into the early 1980s (Santana 1994). By that time, the urban population of the country had tripled from just under one million inhabitants to nearly three million, representing half of the country's total population, one that was over-whelmingly concentrated in the capital city of Santo Domingo (Lozano 2001).[3] As in much of Latin America, rapid urbanization and a weak industrial wage sector led to "informalization" of the urban labor market, characterized by underemployment, unemployment, and precar-ious forms of self-employment – that is, informal commercial services (Lozano 1993; Pérez Sáinz 2003). Over the same period, the number of *minifundios* declined precipitously and the amount of land held by large property owners increased by one-third (Lozano 2001).

An additional factor in the decline of *minifundios* was the surge in international migration. State instability in the post-Trujillo period and US fears of "another Cuba" led to a pro-active immigration policy towards Dominicans in the 1960s and early 1970s as the United States distributed thousands of visas to political dissidents and economic migrants (Castro & Boswell 2002). While the land- and capital-poor moved to Dominican cities, a substantial proportion of smallholding peasants left for New York by leveraging their property for credit to secure their emigration (Georges 1990; Grasmuck & Pessar 1991). As a result, Cibaeño *campos* became transnational sites, receiving remittances and investment from overseas migrants as the ranks of the latter grew.

The social and cultural process of transnationalization of Cibaeño *campos* resulted not only from the out-migration of Dominicans to the United States and, later, Europe, but also from the generalization of Haitian labor in agriculture. By the late 1970s, as the ISI regime foundered and the sugar crisis deepened (see Chapter 2), the sociospatial divisions of agrarian labor, and the related racialization of work instituted at the turn of the twentieth century, were reconstituted in urban and rural spaces. Haitians, formerly restricted to labor on sugar plantations, were incorporated widely into the non-sugar agrarian economy, working as seasonal and permanent laborers in coffee and rice production, widely integrating into Cibaeño agriculture for the first time (Lozano 2001).[4] The incorporation of Haitian agricultural labor did not respond to a functionalist replacement of the male Dominican worker that had emigrated to the city and to the United States over the previous decade, however. In fact, high rates of Dominican male under- and unemployment persisted in rural areas (Dore y Cabral 1981; Grasmuck 1982; Lozano 2001). As I discuss below, practices of self-making and distinction in the *campo,* especially for Dominican men, were tied up with creating social distance from Haitian agricultural workers who lived and worked there.

Gender relations in Cibao *campos* were profoundly transformed through these processes of out-migration and transnationalization as well. In the Caribbean context, as in the United States, women of African descent have always engaged in paid work in significantly higher numbers than women racialized as white. In general, black and mulatto women contributed to their households through wage and non-wage work where the earnings of racialized men were less than a family wage (Safa 1995b; Freeman 2000). Nonetheless, in the Hispanic Caribbean, the white elite patriarchal ideal of a male breadwinner and an unpaid, married female spouse predominated (and continues to do so) despite the actual composition of Dominican households (Safa 1995b). In rural areas like the Cibao *campo,* the patriarchal household remained dominant, but this reality did not mean that women did not contribute to household production and income; more often, women's contributions were not counted by the state, and income from their farm labor was rarely controlled by them (Rocheleau & Ross 1995; Raynolds 1998; 2002). Through the erasures of women's work and the racialized ideal of the housewife, agricultural production for the domestic and export markets in the Dominican Republic was constructed as a male domain. Where households could afford it, women's activities were confined to the home and the all-important *patio* [yard] where women have traditionally cultivated food for household subsistence, except during periods of peak labor demand such as harvests when they have been incorporated into cash crop production. In land- and capital-poor households, however, women sought off-farm work, and with the introduction of new agricultural exports such as horticulture in the 1980s, women's participation in agricultural labor markets increased (Raynolds 1998; 2002).

In the breach between the ideal of women's rural labor as exclusively domestic and the reality of patriarchal households where women's contribution to household income went unrecognized, women have overwhelmingly sought wage work in Dominican cities and abroad. International female migration gained pace in the 1980s, constituting nearly half the migrant stream from the Cibao (Grasmuck & Pessar 1991). Remittance incomes and production strategies of land-holding households with members who had emigrated increased class polarization in Cibaeño *campos* and hastened the pace of rural proletarianization (Grasmuck & Pessar 1991; Ravelo & del Rosario 1986). As a result, the gender ideology restricting women's activities to the home and *patio*, following the white/*mestizo* elite patriarchal ideal, did become a reality in some remittance-receiving households (Grasmuck & Pessar 1991; Ravelo & del Rosario 1986). Remittances allowed women in these households, a relatively small number, to work exclusively in unpaid household labor and this shift increased the social standing of their households. Simultaneously, growing numbers of women from land-and/or capital-poor households – that is, those without the resources to migrate internationally – sought paid work as domestic workers, often in remittance-receiving households in urban and rural areas, as well as in the trade zones (Georges 1990; Safa 1995b).

As we have seen in previous chapters, trade zones proliferated in secondary cities and towns in the 1980s, especially after structural adjustment and the currency devaluation of 1984. In fact, this new export sector mildly attenuated out-migration from the Cibao region and propelled considerable population growth in Santiago (Santana 1994; Ariza 2000). The decentralization of urban growth from the capital city associated with trade zone expansion was due to an increase in migrants, especially young, single women and women in consensual unions,[5] moving to secondary towns and cities to work in these new factories. The participation of women as waged workers in trade zones intensified material and affective connections between rural *campos*, secondary towns, and the region's largest city, Santiago. In addition to remittances, care work was stretched between *campos* and the city as Dominican women who migrated to trade zones entrusted their children to maternal or paternal grandmothers in the *campo* (Safa 1995b). The change in export strategy was accompanied by transformations in household relations and in the composition of households (Safa 1990; 1995b; Ariza 2000; 2004; Itzigsohn 2000). Women's incorporation into paid work was associated with households centered upon the mother–child bond and strong relations among female kin. As rural women abandoned the patriarchal household of the *campo*, and as Dominican men faced declining economic status in both urban and rural labor markets, female-headed households and households with stable, conjugal unions but strong "matrifocal" patterns became more predominant.[6] While female-headed households faced the highest rates of poverty, numerous in-depth studies of female headship have revealed that many women in the new export economies of the Caribbean sought independent strategies to navigate this economic hardship rather than seeking a male partner who would likely be unable to guarantee their economic security.[7]

The city of Santiago had long concentrated surpluses of capital from the commercialization of tobacco, coffee, and cacao grown in the surrounding agricultural region, while remaining relatively small in terms of number of inhabitants (San Miguel 1997). By 1990, although the region as a whole continued to register high numbers of emigrants (surpassing immigrants by half a million), the city of Santiago began to register positive migration inflows as a result of trade zone expansion and associated services (Santana 1994). The city's population doubled over the course of the decade, reaching approximately 600,000 inhabitants. By 2002, 35 percent of trade zone employment in the country was concentrated in the province of Santiago, including the municipality and the surrounding area, and *one out of every six* workers who worked in formal sector employment worked in the trade zones (PNUD 2008), primarily in the garment industry. During the 1990s, the gender composition of trade zone workers also began to shift as more men sought employment in the zones, comprising a slight majority of the workforce by the end of the decade.[8]

Just as trade zone – and garment – employment was concentrated in Santiago and the surrounding towns, so too were the effects of the sector's restructuring. Over the course of the 2000s, export garment sector employment nationwide fell by nearly 80,000 jobs to 42,000. From 2004 to 2006, the period of initial exposure to global competition following the end of multilateral quota restrictions, the number of jobs in Santiago's main trade zone shrunk by over half and continued to decline for the rest of the decade, accounting for 28 percent of trade zone job loss nationally.[9] A USAID-funded survey of 1,500 laid-off garment workers found that unemployment was profoundly feminized across regions, with women experiencing a significantly higher unemployment rate of 24.7 percent (compared to 16.7 percent for men), rising to 48 percent (versus 24.5 percent for men) when women not actively looking for work were included (2007: 72). Of the former trade zone workers who found new jobs, indicators of informality, such as the presence or absence of a labor contract and business license, were high, as were rates of underemployment. Thirty-five percent of former garment workers who had found employment were working less than 40 hours a week and about one-fifth were employed for less than 10 hours (USAID 2007: 51). Women, again, were significantly overrepresented in jobs characterized by informality (USAID 2007: 49). These statistics provide only a partial view, however.[10] Their limits were acknowledged in part by the authors of the USAID-sponsored study. The method of the survey, based on random sampling techniques in working-class neighborhoods near trade zones, was unable to account for those former workers who had left cities and towns where trade zone employment had collapsed (USAID 2007: 46). The gendering of unemployment, as we will see below, must be understood as a spatial process shaped by gendered and racialized livelihood strategies forged through rural–urban connections.

In Santiago, as the pace of factory closures quickened in the mid-2000s, an elite discourse emerged that emphasized the increase in informal activities and framed these activities as both a moral hazard and as a potential danger to the remaining trade zone producers and to social stability more broadly. During the layoffs, for example, the trade zone's management reinforced the zone's perimeter, adding extra rows of cement block to the perimeter wall, replacing barbed wire with circular razor wire, and sealing two of the zone's five entrances.

The director of the Santiago trade zone association explained the need to reinforce the zone's perimeter as follows:

> You would think that with the [factory] closures at the zone, the informal sector activity would go down but instead it has increased. There [the street next to a door recently sealed by the trade zone administration], they sell everything. Not just food [for workers] but women, drugs, everything.

The perception that the increase in informal activity marked a proliferation of abject subjects (i.e., drug dealers and sex workers) was linked not only to the spaces and neighborhoods around the trade zone but also to the problem of migrants seen to be out of place in the city. The mayor of Santiago, for example, declared the following:

> The economic and moral problem of Santiago is that so many trade zone companies have had to close. This has brought a big problem to Santiago because [it is] a city of 750,000 inhabitants, but with 65% of them coming from other towns, people who live in Santiago and worked in the trade zone but who haven't left regrettably ... There are stands and stands on every corner ... And there are also some who we know are already in jail because they have turned to *delincuencia* [criminal activity]. (Fernández 2007)

The mayor framed the employment crisis as largely a migrant crisis, desiring the city to be a conditional residence for migrants. In his narrative, informal economic activity, and especially the very visible practice of street vending, did not index a problem of urban unemployment, but rather a problem of intransigent migrants whose failure to respond "rationally" to their subject position – that is, to migrate – constructed them as potential criminals. Moreover, unregulated work was framed as an exclusive activity of migrants irrespective of informal workers' places of birth or the number of years they had lived in the city.

The criminalization of working-class residents in the city, and the surveillance and control of neighborhoods where they lived, certainly did not begin with the employment crisis in Santiago. Former *zoneros* often mentioned police harassment as a regular part of life in the neighborhoods surrounding the trade zone. Santana, 36, was from a town near the Haitian border and had worked in three different factories in finishing and laundry over 10 years. One night, he recalled, the police arrested him in a neighborhood raid on his way home from the night shift. At the police station, Santana met a lieutenant whom he knew. The officer was surprised to see him at the station and asked why he was there. Santana explained that he had been arrested on his way home from work. "Look," he told the lieutenant, "I still have my [factory] ID card on." The lieutenant yelled at the arresting officers, according to Santana – "Can't you see! He is coming home from work!" – and ordered them to take Santana straight home. Santana mentioned another raid, again at night, while playing dominos on the sidewalk just outside his home to avoid the stifling heat during one of the neighborhood's frequent blackouts. That time he avoided arrest altogether by showing his factory ID.

Raids in poor and working-class urban neighborhoods had increased markedly in 2006 as the Dominican president, Leonel Fernández, implemented a new program of urban policing called *Barrio Seguro* (safe neighborhood).[11] In the narratives of several former *zoneros*, the raids were an opportunity for police to demand bribes from residents who were not carrying identity papers in exchange for not being arrested. Arrest for lack of identity papers has been a long-time, racialized practice in Dominican cities since the target of these random arrests are generally Haitians, Dominico-Haitians, and Dominican rural migrants, many of whom have legal status, but do not possess the means to document their status (Howard 2009). Thus, *Barrio Seguro* intensified, but certainly did not inaugurate, this practice. Santana's story makes a related claim echoed by other workers who told me about the raids. The police lieutenant's reprimand of his lower-ranking officers (*Can't you see!*) clearly reproached them for misrecognizing Santana. The arresting officers had exposed the biases of the state against residents of working-class neighborhoods. Santana's position as a resident in a working-class, immigrant neighborhood – and his provenance as a racialized *rayano,* or migrant from the border region – greatly increased the risk of his criminalization by the state. Nonetheless, using their factory IDs, Santana and his fellow *zoneros* could strategically deploy their status as workers, not to avoid their criminalization entirely, but at least to defer their construction as *delincuentes* (criminals) until the next raid. We can surmise that possessing an ID card of a closed factory would no longer serve this purpose. As Steven Gregory has powerfully argued, the difference between formal and informal labor in these instances was "distinguished less by wages and benefits than by illicit and licit activities, and importantly, laboring bodies that were stigmatized versus those that were not" (2007: 34). Unemployment and informal livelihood strategies increased the exposure of unemployed *zoneros* to detention and arrest, and to having to pay bribes. Several former garment workers whom I would later spend time with in their *campos* expressed relief that they no longer had to deal with the neighborhood raids.

The impacts of both unemployment and out-migration were profoundly felt by residents of working-class neighborhoods who described a significant exodus of migrant workers to hometowns or to tourist poles in the east. For example, they pointed to the many *pensiones* for sale. These were buildings with one-room apartments rented by the week to trade zone workers. Luis, a 22-year resident of the Cienfuegos neighborhood next to the trade zone in Santiago and a staff member of a local job retraining center, described the multiple livelihood strategies he observed:

> There are some, I know a few *zoneros* who have stands to sell *chinas* [cut fruit] on the corner. Today I saw two this morning, one had been a supervisor. But those are just the ones I know. There are few jobs. A big proportion have left to their place of origin, to Dajabón, to San Juan … [Before] in order to leave the neighborhood, you had to wake up very early because it was a terrible mess and you would see lines and lines of people. The *carros* [collective taxis] were full and you couldn't get one. Now, no, you can leave at any hour in the morning and the *carros* are empty. That's the unemployment.

Luis emphasized the emptiness and relative quiet of the neighborhood, characterizing unemployment in terms of an absence of people in contrast to the mayor's complaint of the out-of-place migrants in the street.

While discourses criminalizing unemployed workers point to the risky practices that women and men faced in positioning their laboring bodies after trade zones in the city, rural return was also a prospect that was conditioned by abject subject positions. Below, I explore how unemployed migrant workers negotiated the possibilities for livelihood between spaces and subject positions in the city and in their hometowns. Women's and men's discourses and practices of livelihood after trade zones communicated a resistance to rural return framed around distancing their bodies from abject subjectivities that they associated with the *campo*. Male migrants depicted the risk of being reinscribed as de-valued, racialized rural labor, while women migrants resisted their re-inscription in dependent household relations in their *campos,* which would construct them as unpaid household labor. While both women and men navigated the racial and gendered geographies of the region, women's narratives emphasized the gendered axis of their decision-making. Readers should keep in mind that these gendered discourses of former female trade zone workers were implicitly racialized since, as I have discussed here, household structures and labor market prospects in rural and urban areas were shaped by racialized ideals and practices of marginalization. Taken together, men's and women's discourses of rural return and observations of livelihood practices of return migrants in their hometowns shed light on how unemployed workers negotiated their modern subjectivity in relation to differently gendered and racialized places.

## Workers' Strategies as Embodied Negotiations

The closure of IA manufacturing provided a unique opportunity to study the livelihood practices of former trade zone workers. The presence of two local unions affiliated to two independent federations facilitated my ability to contact and follow workers over time with the help of local union leaders and organizers. Over a period of six months, I spent time with 38 former IA Manufacturing workers – 20 men and 18 women – in Santiago, and in three *campos* in the northern Cibao region. The average age of the workers I interviewed was 32 and the average tenure in the garment sector was nine years working in two to three factories. The Santiago export garment industry had been expanding for nearly three decades by the time of my interviews, and many workers I spent time with had spent the better part of their working lives in garment production, albeit often discontinuously and in multiple factories.

My conversations with many of these workers evolved as I visited their homes in Santiago multiple times, and eventually spent time with a smaller number in their *campos.* Workers reflected on their tenure in the trade zones often with a mix of nostalgia and bitterness. The factory's closure had left them not simply without a job but also without the sociality of employment, forged through department, migrant, and union networks that intersected their place of work. The closure also eliminated the non-pecuniary benefits of the job. The latter were few. The most notable, cited often by women in particular, was access to the company's pharmacy and grocery store where items could be bought on credit at what workers claimed were reasonable prices.

Workers in their thirties and forties expressed significant bitterness towards the company and the sector, often concretely linked to discussion of their severance pay. In most Latin American countries, including the Dominican Republic, no system of unemployment insurance exists. Instead, companies pay a penalty for the right to fire workers in the form of a relatively high severance payment, the amount of which is prorated by wage and seniority (Bensusán 2007).[12] Labor rights activists and unions have long pushed for regulatory changes since severance pay is ultimately managed by employers and can be easily manipulated in their favor.[13] Moreover, the Labor Ministry is often perceived to support employers in disputes.[14] For trade zone workers, severance pay was not solely a legal right. For many, severance pay represented their best hope to accrue savings that they could access as working capital for a livelihood strategy after the factory. In reality, IA Manufacturing workers in most cases used their severance money to pay down debts or for immediate consumption precisely because their regular wages had never adequately met their needs.

Workers experienced their severance pay as the expression of the worth that the company attached to their years of service. Novita, a 13-year operator at the plant in her early forties, described herself as a model employee who had always shown up to work on time, performed her specialized operation with speed and skill, and never called in sick. "Lord!" she exclaimed in reference to her severance, "how well I behaved and how little money I got!" Novita experienced the payment as an insult, and expressed profound regret for her loyalty and good behavior. When she went to the Labor Ministry to contest the amount, the company lawyer discounted her complaint, a decision that was supported by the labor official on duty.

Veteran workers' narratives of regret and resentment were inextricably linked to the feeling that they had sacrificed their youth to the garment sector and received little in return. José, a 38-year-old operator, expressed the sentiment as follows: "In the end, when they threw me out, they gave me 40,000 and some pesos, that's what they gave me and sent me home. I was there for seventeen years ... They took my youth, my strength, everything. It's like [sugar] cane. You take the juice and you're left with the pulp." Like José, many workers experienced their tenure in the trade zones as a kind of theft that had reproduced them as waste. The extraction of surplus value from their labor power had rendered them unskilled bodies wasted through their everyday exploitation in the labor process.

Without unemployment benefits, and with minimal severance pay, former *zoneros* faced economic pressures to find income-earning possibilities. Their strategies, however, were not reducible to economistic logics of labor market insertion; workers' livelihood strategies were simultaneously strategies to forge positions of *social worth*. Women and men sought to position their working bodies not only as the saleable commodity of labor power, but also as embodiments of cultural meanings of progress and modernity. In what follows, I offer four vignettes of worker livelihood strategies organized around workers' narratives and practices surrounding rural return, and the remaking of households that were stretched and strained by the differently gendered and racialized possibilities for constituting livelihoods of social worth.

## Masculinity, modernity, and holding on to "whiteness" in the city

Ramón, Andri, and Hector are each originally from different *campos* in the central Cibao, while Monica is from a *campo* in the eastern Cibao. Monica introduced me to her male co-workers. We first spoke two weeks after the factory had closed in the small yard in front of Monica's two-room *pensión* near the Santiago trade zone that she shared with her husband, also a former IA worker, and their son. Ramón, 26, is the father of three. He worked at IA for 10 years and his last position in the plant was as a *utíliti* (a sub), filling any operation needed on the production line. Ramón explained his desire to stay in the city:

> I have few friends left [in the *campo*] and it's expensive to travel with my kids. Anyway, I have all the relaxation I need with them here. I don't need to go to the *campo* for that. The mentality there is different. [MW: In what way?] In the city, everything is cleaner. You use good clothes and they are clean. In the morning, you shower and put on clean clothes but in the *campo* you get up and put on muddy clothes from the day before and go to work ... My grandmother still lives there with five of my uncles. My family had some land and used to grow cacao, coffee, plantains, oranges ... Now, maybe they have a hectare or two but nothing is being cultivated. You can also study in the city while in the *campo* you can only finish eighth grade. I finished seventh grade before I left and took eighth grade here. I got to the third year of high school. But it was difficult because companies won't respect your study schedule. I'm not going back to the *campo*. I don't like the idea.

Like many other workers of rural provenance, Ramón's initial expectation was to work in the trade zone while completing his studies, an ambition that was thwarted by the long hours required in the garment sector. Monica decided to expand on Ramón's point: "He was *prieto* [dark] and now he is *rubio* [blonde]." Monica used the word *prieto*, a disparaging term for black phenotype often associated with

Dominican anti-Haitianism. Ramón agreed. "Of course," he said, "I have a different color [*tengo otro color*]."

The other two men nodded in agreement. Andri, 26, continued. His older sister had brought him to the city after finishing the eighth grade. His *campo* was near La Vega where his mother still lived, whom he visited every couple of months. Like Ramón, Andri came to Santiago to continue school and stopped studying once his sister got him the job at IA Manufacturing. He worked there for the next eight years, also becoming a *utíliti*. He and the other two men agreed emphatically that there was no going back to *chapeando* (clearing ground) with a machete. Hector, 36, was from Alta Mira, close to Santiago, and had worked in trade zones for 14 years and in IA for the past 11. His last job at the plant was in finishing, attaching cardboard labels to the beltline of pants. He had a small plot in his *campo* not far from Santiago and traveled there frequently. He had worked with horses and livestock there, never with a machete, he emphasized, in reference to Andri's comment about manual rural labor as a kind of going backwards. The livestock had long been sold and he said that he could not return to working with a machete since he had never done so.

Ramón, Andri, and Hector's narratives devalued work in the *campo* and associated rural labor with a lack of progress. Their ascent to connecting this narrative with Monica's description of social whitening suggests that race, class, and masculinity were open to resignification by the movement of their working bodies, a movement spawned by structural changes in capitalist accumulation, but not solely determined by them. Ramón's depiction of the *campo* as a place for relaxation rather than one of labor created further distance between his livelihood strategy in Santiago and the possibility of return to his home village. The exchange between these former garment workers provides a partial perspective into how class is lived as an embodied social and spatial position crosscut with gendered and racial ideologies of labor (cf. Hall 1980).

Historian Silvio Torres-Saillant has proposed an understanding of contemporary Dominican working-class identity in relation to the colonial and postcolonial period. He argues that the distance between social blackness and phenotypic blackness permits Dominicans of color to "step outside the sphere of their blackness, [enabling] them to remain whole" in the face of state negrophobia (Torres-Saillant 1998: 136). Following Torres-Saillant, I want to suggest that narratives like that of Ramón, Hector, and Andri, combined with practices in the *campo* that I discuss below, were part of a daily performance of modern subjectivity in part conditioned by the struggle of Dominican working-class men to resist being inscribed as socially black. Social whitening was achieved through a combination of living in urban spaces, taking on an urban look, and engaging in forms of labor socially constructed as "modern," which these workers associated with factory work. Yet, Ramón's response suggests that the resignification of race through these embodied practices was itself never stable: having another color (*tengo otro color*) is a way of expressing a temporary state – like being cold or being sleepy. As we saw with the neighborhood raids, the factory's closure opened the possibility of Ramón and his co-workers to be resignified as socially black.

## Multiple strategies of men in the campo

While Ramón, Andri, and Hector contrasted their desires for urban modernity against manual, agrarian labor in the *campo*, the majority of return male workers I met were working in off-farm, commercial activities. In fact, of the 10 men I interviewed who had returned to their *campos* in the five months following the factory's closure, only one was engaged full-time in farming. Some men were taking on odd jobs and subsisting on the largely reciprocal and gift economies of their home communities until they might migrate again. This was popularly referred to as *picando*, or nibbling, a way to sustain oneself without getting ahead. In rural areas, this was a decidedly masculinized livelihood form, reflecting the more restricted mobility of women in rural spaces that I discuss below, as well as the masculine construction of paid work.

Some former male trade zone workers had accumulated enough money to allow them to integrate into informal commercial circuits in their *campos*. These men had continued to cultivate their rural social networks during their years in the Santiago trade zone, which now facilitated their return. Moreover, these rural commercial activities were viable because of in-flows of remittances in those communities with a high proportion of international migrants, as well as the relatively lucrative agricultural production of the Cibao, especially domestic rice and export cacao in the *campos* where I spent time.

Marcos, 38, worked in the trade zones for nearly 20 years, starting when he was 18. Despite his long tenure in a total of six different garment factories, he never fully left his *campo* of Rio Grande, just north of Santiago. He built a small house for himself there, which his family members maintained, and he inherited land when his parents died, which he rented to others to farm. The *campo*, for Marcos, was both a kind of refuge and a long-term livelihood strategy. On four different occasions during his years working in Santiago, he had returned to his *campo* for short periods, twice because he was laid off, and twice when he decided he needed to take a break. Once in the mid-1990s, for example, the factory where he had worked for eight years stitching collars on t-shirts increased the workload, which Marcos associated with pressures to change the production system from individual piecework to group-based incentives. Marcos felt that the money he was making no longer compensated his work (*no me rendían los pesos*) and he went to Rio Grande for a few months. When he returned to the trade zone this last time to work at IA Manufacturing as a *utíliti*, his second stint at the company, he went with a clear goal: to earn severance pay in order to put a down payment on a motorcycle so that he could work as a *moto-conchista*, or motorcycle taxi, in his *campo*. "I tried to behave as well as possible so that I could get them to fire me," he explained. In the logics of poorly regulated wage work, Marcos experienced severance pay as a favor that he could earn from a benevolent boss. He succeeded in getting fired in an early round of layoffs at the company and received 10,000 pesos in severance. "I went straight from the factory to the [motorcycle] dealership," he told me. As a *motoconchista* in Rio Grande, Marcos was doing well, especially during harvest times when he could earn two-thirds of what he had earned in a week at IA in a single day. *Motoconchos* were well suited to transport the small quantities of inputs and products associated with small-scale farming along narrow, dirt roads. With his own home, together with income from his land and his new job, Marcos was supporting his wife and her three children. "God willing," he told me, "I won't have to go back to Santiago again."

While elites and the police resignified unemployed *zoneros* who were engaged in unregulated work in the city as migrants and criminals, return migrants engaging in unregulated or illicit trading could circumvent criminalization because of their social position in their *campos*. Miguel, 44, returned to his wife's *campo* where he had forged close ties during his 12 years in the Santiago trade zone, the last six in finishing at IA Manufacturing. He was working for a wood-processing workshop in the back of his in-laws' house. Miguel's income came from trading in trees and finding buyers. "It is called contraband," he explained. "If the police catch you here, you have to pay them a few hundred pesos," he continued, "but if they catch you up in the mountain, they make you plant 100 trees and charge a big fine." While Miguel's informal commercial activities were illicit, the stigmatization of Miguel's labor as informal or illegal by the state was lessened by his sociospatial position as a local in his *campo*. There, he would be subject to lenient penalties and could rely upon his rural social ties to support his ability to work in the illicit tree trade.

These narratives and activities illustrate how former trade zone workers' embodied negotiations of livelihood were linked to malleable constructions of social blackness, constructions that were associated with kinds of work and particular gendered and racial politics of labor in rural and urban places. Part of these workers' struggle was to find livelihood options that positioned their male, working bodies on the modern end of a potentially punitive and inextricably gendered and racialized spectrum of labor. While Ramón, Andri, and Hector sought to avoid the abject social position of rural laborer by staying in the city, return male migrants navigated their positions within the *campo* in multiple ways and constructed alternative narratives of progress and modernity. One afternoon, chatting with a young return migrant named Nelson, he summarized his decision to move back to his *campo* as follows. "You know," he said, "work in the trade zones is only to keep yourself clean in town, not to actually make any money." Nelson's comment reversed Ramón's reasoning in Santiago by identifying an urban look, and the kind of consumption that it required, as a kind of trap. Nelson, and other men whom I met in rural villages, presented such counter-narratives to refute the prevailing idea that they were men who had "gone backwards." Indeed, neither staying in Santiago nor returning to home villages guaranteed positions of social worth. Rather, gendered and racial constructions of labor intersecting in each locality, and structured by political economic conditions, constituted the spectrum that these workers navigated to produce themselves as modern subjects.

## "There's no work for women there": Reluctant return

As we have seen, the embodied negotiations of former garment workers are inseparable from the gendering of productive and reproductive labor. This process of gendering and its variation across villages, towns, and cities emerged as the primary narrative through which female garment workers described their livelihood possibilities after trade zones. In this section, I focus on one woman, Yesenia, from a *campo* near Nagua in the eastern Cibao, and her considerations and experience of rural return in relation to her possibilities for sexual and economic independence. While the narratives of former male garment workers' centered upon the cultural politics and economic prospects for their paid labor in their *campos*, women's narratives focused upon gendered constructions of paid work in their *campos* and related household hierarchies. For former female garment workers, to return to their *campo* was to risk being reproduced as unpaid household labor embedded within constraining and hierarchical gendered family relations.

Yesenia, 26, worked for eight years in the Santiago trade zone, and spent the last five at IA working as a final auditor while she finished high school. Her mother cared for her four-year-old son in her *campo*. I first met Yesenia in her small, one-room *pensión* a few days after IA had closed down. Yesenia's narrative about her livelihood choices revealed her efforts to exert control over how her body was sexualized and gendered, intimately linked to her desire to be a man's primary partner. Yesenia's hopes for the latter were pinned on her boyfriend who had moved to New York several months before. She told me that since he had left, she was trying to be very careful in her neighborhood not to be seen talking to men or out in the "wrong" places, and to make sure she was picking up her cell phone in case he called to check in on her. She was determined to avoid any activities that would spark rumors, which would cast doubt on her fidelity. Her boyfriend was supposed to come for a visit in May and she was hoping he would have a proposal for a shared life together. Yesenia was reluctant, however, to try another emigration attempt on a *yola*, the boats that traffickers use to carry undocumented migrants across the notoriously treacherous passage to Puerto Rico. On her only emigration attempt several year before, her boat was intercepted by Dominican authorities and, in retrospect, Yesenia was sure they had saved her life.

Yesenia shared various reasons for not wanting to return to her *campo*, a narrative that was constantly evolving as she adapted to, and struggled to rework, her livelihood options, and eventually to return for what she hoped would be a temporary stay. She first explained her reasons for resisting rural return in terms of losing an urban look, similar to her male counterparts; but these concerns also extended to her body's shape and weight, tied up with gendered norms of attractiveness. "There, I eat too much," she explained, "and with my body type, I gain weight." Yesenia continued by explaining her preoccupations about what she would do in her *campo*:

> I am not very open to the idea of living in the *campo*. I left when I was 11, first to San Francisco [Macorís, the provincial capital] and then to Santo Domingo and then Santiago. But all the same, there's money in the *campo*, a lot of rice, farming, coffee, cacao. In Nagua [the nearest town], there's a restaurant and a car wash now. I prefer to go for brief stays, maybe 15 or 20 days. But to sit around and not work … There's no work for women there. I had a stand once where I sold chicken, salad and root vegetables. I had good clients and whenever I am back they ask if I plan to start up the business again. If I go back, I will, but I got sick working with the chickens.

Although Yesenia presented her migration from her *campo* in a step-wise fashion, first to the provincial capital then to the capital city and Santiago, she had clearly returned on previous occasions and maintained close links. Her possibilities for rural return were unique among the former female garment workers whom I met, linked to the location of her mother's house in the center of her *campo*, and what she described as a dynamic economic context. Yesenia's father was a large landowner and rice producer, a former two-term mayor and the father of 26 children from several different partners. She was the daughter of his first wife, and this limited the kind of economic support she would receive from him. Nonetheless, her father supported all his independent children with supplies of rice to last the year, which significantly defrayed Yesenia's food costs. Her *campo* was located at the crossroads of the Dominican Republic's changing accumulation model. The relative economic dynamism that she described was linked not only to extensive domestic rice production in the region, but also to the predominance of remittance incomes in the community, as well as a new highway being built through the area to facilitate travel to the new tourist pole of Samaná in the north. The highway project, tourist travel, sustained remittances, and rice production expanded commercial and service activities. Work in these areas, however, was almost entirely gendered as male as the growth of these sectors seemed to do little to alter gender ideologies of women's mobility and the appropriateness of women's paid labor in small towns and villages. If Yesenia returned, she would negotiate these norms through her home-based food-vending business, earning an income from the increasingly commodified relations in her *campo*.

Yesenia described her reluctance to return not only due to more strict control of her income and movements by her mother, but also due to the presence of her son's father in her *campo*. She described him as a womanizer. His entire family lived in the United States and he received significant remittances. He used the economic power granted by these funds to maintain a sexual relationship with Yesenia, despite being remarried with three children. Yesenia did not want to be his lover, especially living in such close proximity to his wife and her family, and given her desire to be a primary partner. "If I move back," she explained, "he will make my life difficult."

About three weeks after we met, Yesenia did move back to her *campo* and quickly started up her chicken-selling business again. I visited one month later. Yesenia had built a small, one-room extension on the back of her mother's house while working in Santiago; she now slept there with her son. Her days began at 6:30 in the morning. She would put four pounds of beans to boil and start to slaughter chickens. Yesenia took each chicken to the back of the yard by an irrigation canal, tied the bird's feet to a black rope strung from an orange tree, and slit its throat. By 8:00, Yesenia and her mother were vigorously defeathering the birds, swirling them in a pot over a wood fire to ease the task, and taking short breaks to sell gasoline in Presidente-brand beer bottles to passing motorcycle and scooter drivers, along with sweet breakfast beans. By 9:30, chicken sales would be in full swing. Women stood in line and waited while Yesenia finished cleaning the pieces they had ordered. By 11:30, Yesenia would be out of chicken and the day's paid work would be winding down. One morning, at the height of sales, Yesenia's younger sister stopped by and told her that she should change out of her bike shorts and purple jog bra because her father would be annoyed to see her dressed that way. Yesenia was too busy to act on her sister's advice, but the next day she was wearing long sweatpants and a t-shirt.

Yesenia told me that she earned more money from her business than in the trade zone because she did not incur many expenses.[15] She and her mother kept their accounts separate: the gasoline business was her mother's. And while Yesenia passed her savings from the chicken sales to her mother, she passed the money from the sweet bean sales to one of her sisters who lived down the road to keep aside for her. Yesenia was trying to save in order to make it back to Santiago. The city, for her, was a refuge of sorts from her restricted mobility in her *campo* and the controls and expectations on her behavior maintained by the dense network of kinship relations that surrounded her there. Back in Santiago, she might constrict her movements to make sure not to arouse suspicions of her commitment to her boyfriend in the United States, but she perceived her self-vigilance there to be of her own choosing. This self-policing was clearly less onerous to her than the social limitations placed upon her in the *campo*. Her feelings of independence in Santiago were bolstered by her increased control over her income and spending on her "urban look," even though she had found herself increasingly indebted prior to the closure of the plant. If the trade zone factories re-opened, Yesenia was sure she would return.

## Stretching family ties across uneven geographies

Yesenia's access to steady, paid work that conformed to expectations of women's home-based labor was unique. For the majority of female garment workers, returning to the *campo* would mean the resignification of their labor as unpaid and domestic. Women's livelihood prospects in rural villages and towns contrasted sharply to those of men who could potentially find some work *picando* at a minimum, or working in informal commerce, transportation, or agriculture. This tension between women's and men's place-based work possibilities caused conflicts within households that were adapting to the new reality of unemployment in the wake of the trade zone's employment collapse.

Laya, 26, and Domingo, 38, a couple from a *campo* near Nagua, had both been working at IA Manufacturing since 2000. Laya moved to another garment export factory where she continued to work as a *taqueadora* (cross-stitching belt loops) shortly before IA closed. With the closure of the plant, Domingo lost his job pressing pants in finishing and reluctantly returned to cultivating rice on his father's 60 *tareas* of land (3.75 hectares). Laya remained in Santiago with their two children earning a wage largely to cover her costs of rent and childcare in the city. She was hoping to keep her job in order to accrue sufficient severance pay to buy her own house in the *campo* in order to set up her own business. She was clear and adamant about her strategy, despite Domingo's desire for her immediate return with their children to his parent's house. Laya was resisting Domingo's request. She planned to work for as long as possible in Santiago in order to create a place for her paid labor in her *campo*, as well as independence from her husband and his family.

This particular gendered geography layered on top of the more common arrangement of social reproduction where older family members – usually grandmothers – cared for children while parents worked in the city. With job loss, some unemployed workers were sending additional children to live with family in rural hometowns. The loss of income meant that they could not visit their children with as much frequency, nor send key consumer goods such as clothing and toiletries. Finally, households that had family members who could send remittances from the United States or Europe were becoming more dependent upon these crucial funds and goods that arrived from the north.

Suzana, 33, was a single mother of five children. In 1994, she moved to Santiago with her newborn daughter, leaving her three older children with their father in a *campo* near Azua, in the south. That same year, her mother left for Puerto Rico as an undocumented migrant. Over the next 12 years, Suzana worked in the trade zones in five different factories, ending up at IA Manufacturing for three years. Through her mobility between factories, she trained herself to sew several different operations and returned to her studies, eventually finishing high school. Her oldest daughter came to live with her in Santiago in 2001, but she was only able to bring the other two children, now 16 and 14, from Azua in 2006, shortly before the factory closed. At the time, Suzana was depending on her mother's remittances to support the family. "Things have not gone well for my mother in Puerto Rico," she told me. "She's 62. I'd like to be able to help her out, not the other way around." To support the household, Suzana's mother was sending packs of new clothes from Puerto Rico. Suzana had set up an informal *agachate* boutique[16] in a side room that she built next to her living room where neighbors could peruse the merchandise. The business was floundering since few people could make payments on the clothes, and Suzana was hoping to find work soon, while also exploring avenues to emigrate to Spain. After 12 years in Santiago, however, it was unthinkable to Suzana to return to Azua. "I'm not one of those people who likes to go backwards," she told me. Such a common sentiment, as we have seen, can only be considered in gendered terms. Suzana had found a certain limited freedom to study, to negotiate her labor by moving among operations and between different factories, eventually becoming active in the union, and to organize her own household in Santiago. She was sure that these possibilities would be unavailable to her in the *campo*.

Migrant women's negotiation of their position with respect to unpaid domestic work was certainly not guaranteed by staying in Santiago. Like Suzana, most of the women migrants I interviewed who were staying in the city were looking for paid work with little success. The majority of women were either engaged primarily in unpaid household labor (and dependent on spousal support in cases where they had partners), dependent on remittances, and/or working as domestic workers. Nevertheless, these women argued that to return to their *campos* was to foreclose the possibility of their paid labor.[17] To stay in Santiago then was to position themselves as the embodied potential of paid labor that could be realized through the market, despite the staggering unemployment rates that signaled the failure of this realization as practically the norm. The impossibility of their paid labor in rural villages and towns was not a property of rural areas as static or traditional places, however; rather, as the comparison to men's rural return suggests, the gendering of labor in rural areas was an effect of the way that intimate, regional, and trans-national connections were woven through these localities.

## Conclusion: Producing Workers on the Threshold of the Idled Factory

The geographies of work discussed here call for a rethinking of critical approaches to global production. Feminist ethnographies of global factory floors have meticulously documented the ways in which gender, and especially docile femininity, is produced (or not) through the practices and discourses of management. This approach has challenged the dominant notion that workers with these characteristics are simply hired by recruiting labor from seemingly stagnant and uniform gender contexts. In so doing, the goal of feminist scholarship has been to evacuate the essentialist core of the Third World woman worker as perpetuated by capitalist discourses.[18] But as my study of factory flight demonstrates, subject production is not reducible to workplace practices, especially in contexts where workers must develop strategies to navigate unstable and low-paid wage work. In short, while we can agree that working subjects are produced through the labor process, thus contesting fixed, often racist and sexist notions of cultural traits, this view unfortunately erases the unstable geography of global production itself. In the context of factory flight and ephemeral wage relations, the production of subjects may take place as much, if not more, through the very instabilities of the capitalist labor process itself.

By examining women's and men's experiences after garment work in the Cibao, the challenges and possibilities for analyzing complex subject positions and forms of subject-making in relation to capital accumulation and regional formation come into sharp relief. For men of rural provenance, keeping social and spatial distance from agrarian labor constructed primarily as an abject position through an exclusionary axis of race, class, and nation (i.e., black, proletarianized, and Haitian) is one manifestation of embodied negotiations. This strategy leads to multiple practices depending on migrant men's connections to and possibilities for livelihood within their *campos*. In the articulation of race, class, and gender experienced by Dominican women, the gendering of labor appears as a dominant relation although inseparable from existing race/class hierarchies. The white/*mestizo* patriarchal ideal of the male-headed household, one that has always materialized unevenly in the Dominican context, has undergone contradictory changes through the transnationalization of *campos* and the incorporation and expulsion of women into and from paid work. Nevertheless, former female garment workers' narratives suggest that this ideal household structure continues to gender paid labor in *campos* as male. Staying in the city thus remains a key strategy to stay in circulation as potential paid labor, a strategy woven through women's efforts and desires to embody positions of independence that are inseparably economic, social, and cultural.

At the threshold of the idled garment factory, these multiple practices of subject-making come to the fore to reveal complex geographies of work. Workers navigate the unstable boundary between wage and non-wage work by mobilizing livelihood strategies and cultural meanings of progress to forge subject positions that offer distinction from lesser, abject social locations. In short, workers experience the fundamental dialectic of value and waste that makes capitalist accumulation possible as a variety of sociospatial strategies to construct positions of social worth. What workers' narratives and their strategies reveal is that this process is on-going. The boundary between wage and non-wage work is not hard and fast; workers negotiate this limit under conditions not of their own choosing. *Zoneros* navigate the instabilities of wage work by crafting alternative discourses of progress and the possibilities for work outside of the wage relation. They do so even as they risk being reinscribed as waste and non-value by capitalist logics shot through with racialized legacies of colonialism.

In the following chapter, I turn to the process of global factory expansion on the Haitian border where Dominican and US capital constructed a new trade zone in the early 2000s. We will see that despite the seemingly incontrovertible logic of bringing thousands of jobs to a place with high unemployment, the efforts of transnational capital faced numerous overflows, contestations, and alternative practices of meaning-making by workers. New Haitian garment workers, like their veteran Dominican counterparts, sought collective and individual strategies to forge positions of social worth while navigating both the exploitation of the wage relation as well as its volatility. The chapter shows how uneven development is made through reifying the wage/non-wage boundary by carving out a border trade zone and the challenges to this idealized geography by workers who circulate in and through the global factory.

## Notes

1   This chapter is a revised version of Werner (2010). Republished material is printed with permission from Taylor and Francis. www.tandfonline.com/doi/full/10.1080/0966369X.2010.517023

2   The ethnographic shift that I am pointing to here responds to the contemporary conditions of global production. Consider that many of the insights from Burawoy's groundbreaking ethnography were made possible, he tells us, by the fact that the factory where he worked had been the subject of a previous study some 30 years earlier. It is safe to say that such a "revisit ethnography" would be impossible in today's global factories. In fact, the factories that are the basis of many widely read studies on global production are oftentimes closed by the time the monographs are published. Despite this fact, few authors incorporate factory closure into their studies (although see Collins 2003).

3   The ratio of population between Santo Domingo and Santiago increased from 2:1 in 1935, to 3.2:1 in 1950 to 4.1:1 in 1981 (Santana 1994). This trend in urban primacy was typical for the region. For more on Caribbean urbanization trends in a comparative framework, see Portes et al. (1997).

4   Lozano's political economic analysis of rural labor markets and Haitian migrant labor in the region's rice economy remains unparalleled to my knowledge (Lozano 1998; 2001). I know of no in-depth ethnographic studies of Haitian rural agricultural labor in non-sugar production. The small but growing English-language literature on the racialization of labor in the Dominican Republic is conducted largely in urban contexts. Two groundbreaking studies are Howard (2001), based on research primarily in Santo Domingo, and Gregory (2007), based on research in and around the tourist industry in the declining sugar mill town of Bávaro, just east of the capital. A third study takes a more explicitly transnational approach, including ethnographic work from New York City (Candelario 2007).

5   A minority of working-class Dominicans and those of rural provenance are legally married. Below, I use the term "marriage" and associated terms of "husband" and "wife" when informants used these terms to describe their household relationships.

6   Matrifocality is the classic anthropological term used to describe and distinguish patterns of household relations among Afro-descent, urban working classes from the patriarchal, male breadwinner model associated with the white/*mestizo* household ideal. The origins of matrifocality are varied, traced to African retentions, the legacies of slavery, and male migration and marginalization (Safa 1995b: 56). Yet, as Blackwood (2005) notes, such a construction has painted diverse, often multi-generational households with too broad a brush, and has reinforced heteronormative assumptions by centering the idea of the household around an absent heterosexual man, while erasing the possibility of households based around same-sex bonds. I use the term advisedly here to indicate a spectrum of household forms that center around women's decision-making and income-earning.

7   For example, see Deere (1990), Safa (1995b), Freeman (2000), Ariza (2000), and the unpublished trade zone ethnography by Dominican anthropologist Fatima Portorreal (1991). In my conversations with trade zone workers, the weakness of the heterosexual conjugal bond, and the corresponding tensions between women and men, was a frequent topic of conversation. Women in female-headed households often declared their frustration with Dominican men through comments such as "*el hombre dominicano no sirve*" [Dominican men are no good], and pledges never again to share a household with a Dominican man. Safa quotes a trade zone worker from the 1990s as saying "*Yo no doy mis hijos por un hombre, y mucho menos por lo[s] que aparecen hoy*" (I wouldn't give up my children for a man, and much less for those who are around today) (1995b: 117). Men, on the other hand, complained that women only paired up with them as long as they had a job, and broke off the relationship once the men were unemployed. One unemployed male *zonero* commented, for example, that "love comes through the hand" (*el amor entra por la mano*), meaning that love is conditioned upon men providing income to the household.

8   See Chapter 3 for a discussion of the changing gender composition of the trade zone workforce.

9   The northern region, including the Santiago trade zone as well as those zones on the city's periphery and in secondary Cibao towns, concentrated 55 percent of the job loss over this period. (CNZFE annual reports, various years).

10  For 2006, the national unemployment rate was 16.2 percent. In Santiago, male unemployment in 2006 was 8.5 percent versus 20.4 percent for women, a ratio that is slightly more favorable than the national figures that show female unemployment to be three times that of male unemployment (PNUD 2008: 262–263). A general observation about the Dominican labor market repeated by most development institutions including CEPAL, the International Labor Organization, the World Bank, and the UNDP is that

rapid GDP growth has had relatively little impact on employment expansion (Sánchez-Fung 2000; Reyes 2001). Unemployment has never dipped below 12 percent, even when GDP has registered phenomenal growth rates of 10 percent. In gendered terms, economic growth has vastly expanded female unemployment. This trend is apparently not consistent with outcomes across low-income countries where there has been a general convergence between men's and women's increasing unemployment, although the Caribbean maintains higher rates of female versus male unemployment relative to those recorded in Latin America (Standing 1999: especially 597–599). These figures also erase a whole host of fundamental social relations such as remittances, internal and international migration, the politics of visibility and invisibility of Haitian undocumented labor, numerous other livelihood strategies that are either ignored or criminalized, and the very problem of locating "the labor market" in a terrain constituted by relative mobility and immobility. I address only a small subset of these concerns in this chapter.

11  *Barrio Seguro* was the flagship program of the President's Plan for Democratic Security, announced and piloted in one neighborhood in Santo Domingo in 2005. The zero-tolerance police program was then rolled out to 13 neighborhoods in the capital and expanded to a handful of neighborhoods in Santiago in 2006. The program involved the recruitment and training of 16,000 police personnel, and later extended to military personnel, deployed to undertake intense surveillance – including street patrols and group round-ups – of targeted low-income neighborhoods. The rhetoric and execution of the plan was aimed at "containing" crime within low-income neighborhoods constructed as threats to Dominican society. See Howard (2009) for a discussion of the program in Santo Domingo.

12  Because women earned lower average wages than their male counterparts (an issue I address in Chapter 3), women workers also received less severance pay. In a national survey, 48 percent of laid-off women garment workers who received their severance pay received 6,000 pesos or less compared to 37 percent of men (USAID 2007).

13  In 2004, Dominican unions won a legal battle against the trade zone sector when the courts ruled that the widespread practice of paying severance annually was a legal violation. Until that time, trade zone companies were technically re-hiring workers on an annual basis and thus not compensating them for their seniority.

14  In *maquilas* and trade zones throughout the macro-region, it has long been a practice to close a company that has accrued a high indemnity in the form of owed severance illegally, and to transfer all the assets to a secret partner so that workers – who attempt to recover the debt by petitioning the state to place an embargo on the assets – cannot recover their severance. In cases of this method of corporate theft, often tacitly if not openly tolerated by the state, workers have seized company installations through occupations and have used the threat of property destruction. Workers told me about several such occupations in the Dominican Republic during the industry's heyday

although no actual cases of property destruction. In Guatemala in 2007, local media reported that garment workers burned downed the installation of their factory when the owner refused to pay their severance (Ramírez 2007).

15   Yesenia estimated her daily earnings through food vending to be between 200 and 300 pesos after expenses, totaling 1,200–1,800 pesos per week. Yesenia's last base wage in the factory was 1,200 pesos per week. She frequently earned between 1,500 and 2,000 pesos per week with production bonuses. In the last two years of work, Yesenia claimed she was unable to save money because of increasing costs in Santiago.

16   From the word "agacharse" or to bend down, since the clothes were put out in stacks on the floor.

17   Women's reluctance to return to hometowns and cities has been noted in the context of international migration by Grasmuck and Pessar (1991). Women's experiences negotiating gender norms of labor, sexuality, and household hierarchies upon return are also discussed by Gregory (2007: 69–80).

18   Leslie Salzinger's widely read ethnography of *maquila* shop floors on the United States–Mexico border, for example, describes femininity as a trope mobilized by management to interpellate factory workers – that is, to hail workers as the docile feminine ideal of global production. She writes that "[t]he notion of 'productive femininity' thus crystallizes through a process of repetitive citation … Contrary to managerial hopes and feminist fears, docile labor cannot simply be bought; it is produced, or not, in the meaningful practices of shop-floor life" (Salzinger 2003: 16). While I am sympathetic to the project of contesting an essentialist Third World woman, this project remains limited in its overemphasis on dominant, managerial discursive practice, while failing to theorize the more material-discursive nexus of wage work instability, one that is measured but poorly described by the well-over 50 percent turnover rates in many global factories around the world, together with significant factory flight, which remains largely invisible and unregulated (see also Bair 2010).

# 5

# Reworking Coloniality through the Haitian–Dominican Border

*Haiti is not the future. It is the present.*
— Levi's sourcing manager for the Dominican Republic

## Introduction[1]

By the mid-2000s, as Dominican garment workers faced mass retrenchment, the global factory was re-emerging in Haiti.[2] Like the Levi's agent quoted above, garment owners and managers working for companies outsourcing to Haiti emphasized the urgency of incorporating Haitian workers into the garment labor process. To this end, a handful of Dominican garment companies, as well as large North American firms with operations in the Dominican Republic, were searching for business arrangements in Haiti to make this reordering of production possible.[3] They faced innumerable obstacles: from poor roads to get the components and finished products to and from Port-au-Prince, to political instability, including frequent transport strikes, to difficulties in working with an entrenched commercial elite in Haiti's capital who were loath to give up significant shares of profit and control.

Faced with these seemingly insurmountable barriers, the Santiago-based garment giant Dominican Textile (DT) undertook an ambitious plan to incorporate the northern border region into the garment export sector through the construction of the island's first border trade zone.[4] While the company president fervently defended the firm's position as a major employer of Dominican workers in Santiago (see Chapter 3), he,

*Global Displacements: The Making of Uneven Development in the Caribbean*, First Edition.
Marion Werner.
© 2016 John Wiley & Sons, Ltd. Published 2016 by John Wiley & Sons, Ltd.

like his client at Levi's, affirmed an inevitable shift of assembly work to the country's low-wage neighbor. From DT's perspective, the northern Haitian border town of Ouanaminthe, 140 kilometers west of Santiago, was a more than adequate "greenfield" site to build the new trade zone. "We were bringing jobs to a town with 90 percent unemployment," reflected the company president when I asked him about the many protests that had challenged the project. "I believed we would be greeted with a red carpet," he continued, still in disbelief over the contentious reception the company had received. Seeing the residents of Ouanaminthe as a static pool of reserve labor, the company president had apparently failed to understand the town's radical – and on-going – transformation.

After decades of official closure, and the restriction of legal trade, the border's new relative openness, following the collapse of Haiti's dictatorship in 1986, was remaking the region. In the tumultuous decade and a half that had followed, Ouanaminthe had been thrust into the center of Haiti's provisioning network. The town had become a key site for a relatively lucrative exchange between a host of Dominican petty traders and agro-industrialists, mostly from the Cibao region, and Haitian traders, both local and from distant regions of the country. Together with Haitian farmers and migrant workers, these groups were all in a sense harnessing the emerging geographies of unevenness crystallized at the border. Their transnational strategies generated significant returns for many of the Dominicans involved and minimal livelihood possibilities for many Haitians in the context of country's collapsed economy. The reworking of labor and export garment production across this profoundly uneven divide would be both enabled and deeply constrained by these multiple forms of interdependence that were emerging between the two countries.

The border trade zone can be understood as a classic "spatial fix" in the sense that capital sought to construct a new geography of production in order to resolve the dilemma of profitability of the prior arrangement (Harvey 1999 [1982]). And yet the place, Ouanaminthe, and the value hierarchy of Haitian and Dominican labor that investors sought to materialize there, was not merely available to be slotted into the lowest rung of the garment production network. Rather, coloniality and the uneven geographies of production it enabled had to be reproduced. This chapter, then, excavates the conditions of possibility for this new strategy of accumulation and its limits. First, I explain how recent political transformations of the northern border region allowed for its production as a site for accumulation. I take neither the continuum of exploitable and hyper-exploitable labor produced as "national" – that is, Dominican and Haitian – nor the appearance of this apparently new low-wage frontier, for granted. Rather, I show how this social hierarchy was reproduced in new ways through the vicissitudes of political transformations that profoundly reshaped the border region itself. The second task of the chapter is to detail and to interpret the limits of this new arrangement. US and Dominican garment capital and the Haitian state faced organized resistance as well as everyday sorts of interruptions and conflicts with emerging livelihood strategies at the border. These limits were not only internal to the contradictions of capital accumulation; they also emerged from the way in which the capital–coloniality nexus was reproduced between Haiti and the Dominican Republic and materialized through the border region in particular. In short, the efforts of capital and the state to produce new geographies of accumulation faced continual ruptures and instabilities, consolidating the trade zone at a smaller scale than initially envisioned, while generating new politics and new aspirations at the border.

The chapter draws primarily upon interviews with community leaders and labor activists in Ouanaminthe and managers from Dominican Textile, as well as observations inside the border trade zone.[5] I proceed in four main sections. In "Remaking the Border," I detail the transformation of Ouanaminthe in the wake of the political transition from dictatorship and the rise of the Lavalas movement in the 1980s, focusing in particular on the embargo against Haiti initiated in 1991 following the ouster of President Jean-Bertrand Aristide. In the section "Reordering Trade Zones between the Dominican Republic and Haiti," I tease apart the political conjuncture that brought together unlikely allies in the early 2000s – local farmers and landowners, anti-neoliberal activists, and right-wing anti-Aristide campaigners – to organize against the implantation of the trade zone on the border. The fourth section, "CODEVI," details the ensuing struggle for labor rights in the zone in the context of Haitian state collapse. In "Disposability at the Border," based on my observations in the zone, I discuss the daily practices of power – and their limits – that underpinned the zone's strict division of labor between Haitians and Dominicans. The apparent fixity of the trade zone was constituted by its "revolving door" character, as high numbers of managers and workers were either fired or decided to leave, propelled by workplace conflicts and the intensification of unevenness at the border.

In the conclusion, I return to the notion of coloniality and its meaning in the reworking of the border, and the attempts by garment capital to extract surplus value across this divide. Accumulation at the border and the expansion of Dominican exports and transnational foreign investment to the region illustrate how coloniality is harnessed by capital through the intensification and fragmentation of sociospatial unevenness. At the same time, the character of this "spatial fix", and the dynamic geographies of unevenness that constitute it, is not determined by the demands of capital alone. Layered histories of violence, nationalist and democratic movements, and class struggle, together with the gendered livelihood practices that have emerged since the border's opening, continue to shape the geographies of uneven development in contingent ways.

## Remaking the Border: The Haitian Embargo of 1991–1994 and the Growth of Ouanaminthe

The Massacre River that delimits the national border that runs between the towns of Ouanaminthe, Haiti, and Dajabón, Dominican Republic, defines the divide both physically and symbolically. Named for a conflict between French buccaneers and Spanish colonists in the eighteenth century, the river was resignified through the violence of modern statehood when many of the estimated 15–20,000 victims of Dominican President Rafael Trujillo's 1937 ethnic cleansing campaign died there, while survivors crossed its wide bed to safety in Ouanaminthe. The massacre, popularly known as *el corte* (the cut or wound) in Spanish or *kout kouto-a* (the stabbing) in Creole, was directed against Haitian and Haitian-descent peasants in the northwest of the Dominican Republic, especially in the areas surrounding Dajabón. With the important exception of seasonal Haitian migrants who crossed the border to work in the distant Dominican/US sugar plantations of the southeast, the result of the massacre was the formal closure of this erstwhile fluid frontier for nearly five decades.[6]

The sparse record of Dominican state intervention in the region following *el corte* demonstrates that Trujillo's early preoccupation with the border was an aberration. The massacre, followed by efforts to establish settler colonies in the 1950s, remained the only precedents of active state policy in the area (Augelli 1962). Investment promotion incentives (including a trade zone promotion law), adopted throughout the second half of the twentieth century to provide tax breaks to businesses that located in the Dominican border area, generally failed to spur significant investment due to the lack of basic infrastructure, including electricity and schools. For decades, then, much of the population of the Dominican border region emigrated in search of education and jobs to Santiago and even to the distant capital. Many of these border migrants arrived in the Dominican trade zones to work in garment assembly throughout the 1980s and 1990s.[7] Overall, the Dominican state's presence in the border region largely materialized in its military form as various elements of the armed forces policed, and trafficked across, the divide (Silié & Segura 2002).

Similarly, in the neighboring border region in Haiti, the state's primary manifestation was its military: the largest deployment of soldiers outside of Port-au-Prince resided in Ouanaminthe. Following the 1937 massacre, Haitian residents of Ouanaminthe, including many survivors, lived for decades without venturing the short half-mile to Dajabón. For some, this was a political and personal decision taken in the wake of the massacre, although the Haitian state also forbade cross-border traffic, at least under Francois Duvalier.[8] While a clandestine trade existed between the two militaries, other cross-border activities were restricted.

Over the course of the half-century following the massacre, then, Ouanaminthe remained a relatively small town with an expansive agricultural periphery integrated into the social and economic fabric of Haiti's northeastern region. Agricultural products such as mango, yucca, rice, and avocado were brought to the town from the surrounding countryside and some surplus produce was taken by road to Fort Liberté and Cap-Haïtien. The only major investment in the region was a large sisal operation, the Dauphin plantation, in nearby Morne Casse. The plantation operated for 40 years, generating considerable employment in the area, before being abandoned by its US owners when the market for sisal collapsed. Local residents' recollections of Ouanaminthe during the period of the border's closure are deeply nostalgic: a pastoral town of about 10,000–15,000 people based around four main streets where everyone knew one another, free from the trappings of urban vice like bars, crime, and prostitution. In the 1970s, Jean-Claude Duvalier inaugurated a generator, providing between three and six hours of electricity a day, and a water system, both serving those who lived in the town center. The regiment of nearly 600 soldiers stationed in the town included recruits from the northeast region and from the town itself.

Shortly after the fall of the Duvalier regime in 1986, restrictions on cross-border trade were eased and a bi-national market quickly emerged, fueled by Haitian tariff liberalization imposed by the United States and the IMF (Weisbrot 1997; Saintilus 2007). On Mondays and Fridays, local Haitians began to cross the bridge over the Massacre to Dajabón to buy foodstuffs, and on Saturdays and Tuesdays, local Dominicans started coming to Ouanaminthe to purchase cosmetics, used clothes, electronics and, eventually, the US-produced "Miami" rice that was beginning to flood the Haitian market.[9] By the 1990s, sources estimate that the average tariff in the Dominican Republic was 20 percent higher than in Haiti; however, the practice of taxing imports was likely less systematic than such a statistic suggests.[10] Other forms of cooperation emerged based on the widening inequality between the two countries.

For example, Haitian peasants hired Dominican combines for the rice harvest, a service they paid for by giving over a portion of the harvest to the machine's owner.

Ouanaminthe would not be immune to the country's upheavals in the wake of the Lavalas movement that gathered force during the late 1980s and swept its leader, Jean-Bertrand Aristide, into power in 1990. Among other reforms, Aristide abolished the army, a force that had brutally served the interests of Haiti's "parasitical," albeit fractious, political elite (Dupuy 1988; Trouillot 1990). The move further isolated Ouanaminthe from the Haitian state and reinforced, for some residents, their sense of relative powerlessness with respect to the Dominican armed forces whose presence was felt strongly at the border. After only eight months in power, Aristide was ousted by a military coup and the Organization of American States (OAS) called for an embargo in condemnation of the action in 1991.

Under the embargo, Ouanaminthe quickly became the terrestrial port for a significant proportion of oil-based imports from the Dominican Republic, sanctioned by Dominican President Joaquín Balaguer, the intellectual author of modern Dominican anti-Haitianism, and vocal opponent of President Aristide.[11] As the gasoline, diesel, and kerosene traded across the Massacre River began to supplement more and more of Haiti's basic energy needs, a diverse influx of migrants from all over the country arrived in town to participate. Residents described the immigration as akin to a gold rush. The new migrants included wealthy and well-connected traders as well as the land- and capital-poor. These latter migrants, men and women, lived in makeshift shelters on the banks of the river and sold gallons and jugs of gasoline that they carried across to intermediaries. The wealthy traders coming from the main cities transported the fuel to semi-paralyzed industries and businesses in the country's urban centers. Agricultural plots near the river, occupied by temporary shelters, holding tanks, and tanker trucks, became so steeped in gasoline and waste that farmers thought they would never recover the land.

Ouanaminthe residents' relationship to the gasoline trade was ambivalent. Many participated either directly in the trade or in the provision of related services. Several interviewees from the town's elite reflected upon how the trade took over the town by giving the example of school teachers who left their classrooms in the middle of term, borrowed small amounts of money, and quickly achieved an economic status previously unimaginable for their social position – for example, by building their own homes and buying motorcycles. In other words, the gasoline trade led to the accumulation of what Janet Roitman (2003) has called "unsanctioned wealth," or wealth that is not authorized by prevailing social hierarchies and is in fact disruptive of these. As one teacher and member of a prominent Ouanaminthe family explained:

> It was a kind of social and economic explosion … The population was not prepared for this development and it had many consequences. From a cultural perspective, there was a loss of culture. We started to have problems of youth delinquency, epidemics, high insecurity, and depravity of our customs.

A local labor rights lawyer was more sympathetic to the plight of the migrants and the role of residents – both in town and in the surrounding districts – in fostering their social exclusion:

> Still now, people say 'those others who come here.' And if there's a problem, they blame the people who have come here. But there are people from Ouanaminthe … [who] are more accepting … There's discrimination. Not just by people from Ouanaminthe [the town] but also from the surrounding districts [*sections communales*] because the person from Dilaire [a nearby village] for example knows that he is from Ouanaminthe. He also has problems with people from Cap-Haïtien or San Michel. To date, there's a problem with integration. Each year, the population grows because people say in Ouanaminthe there is work, in Ouanaminthe there's trade, you can go to the Dominican Republic, you can do something.

From the perspective of these and other pre-1991 town residents, the embargo had produced a weakening of the town's identity as Ouanaminthe had absorbed migrants from all over Haiti, noted as outsiders by, for example, their religious practices and styles of dress. Such social divisions hampered the efforts to organize community responses to the many challenges that the town faced, as we will see when we come to the struggle over the trade zone.

The trade also fueled a building boom that threatened landholdings, large and small. Some landholders quickly sold their properties to preempt their occupation by landless migrants, while some smallholders managed to rent pieces of their plots and their yards to new arrivals. With the influx of migrants and capital funneled into home building, Ouanaminthe saw the rise of numerous *cités*, new unauthorized neighborhoods often established through organized land occupations. Elites, officials, and community leaders commonly described the town's growth as "anarchic" – that is, as unplanned and unorganized. They contrasted their description with neighboring Dajabón's relatively more planned and regulated urban order. While officials in Dajabón were able to channel some of the capital generated by the boom into the construction of municipal buildings and a hospital, the instability of the state in

Ouanaminthe meant that profit from the new trade was accrued individually and secured largely through the construction of individual housing. Pre-1991 residents' narratives conveyed a mix of sentiments: being besieged by outsiders, crime, and contraband, while benefiting from new opportunities. They identified as and with "native-born Ouanaminthe residents" (*moun wanament natif natal*), a phrase that denoted not simply being born in the area, but rather coming from families who were established there before the embargo.

In effect, the embargo set off a series of events that transformed both Ouanaminthe and Dajabón, and their relative positions in the evolving and inter-related patterns of accumulation and dispossession between the two countries. The embargo's effect on Haiti's productive sectors was devastating, and the subsequent political instability and neocolonial interventions continued to hamper the prospects for (re)construction. After serving as a temporary site for gasoline provisions, Dajabón became a permanent depot supplying a significant portion of Dominican goods to Haiti, and receiving goods transshipped through the latter. The production and distribution of ice and water is a good example. The embargo precipitated the closure of the regional ice factory in Cap Haïtien. Aguas Beler, a small Dajabón-based ice factory, boosted production and diversified. From a daily output of around 80 blocks a day in the early 1990s, by 2006, the operation was selling hundreds of ice blocks to Haiti daily, plus 30–35,000 small water bags per week, as well as distributing propane gas, hundreds of tons of rebar a month, and other building materials. At the time of my research, the company was the only supplier of ice and purified water to Ouanaminthe, employing 50 direct workers.[12]

The embargo effectively transformed the market between the two towns in scale and scope, expanding to more consumer as well as non-consumer goods and becoming a main conduit of supply for the northern region of Haiti, reflecting a new trajectory of unequal inter-dependence that long outlasted the political event (Dilla & de Jesus 2005). The market activity itself was transferred to the Dominican side of the border soon after the embargo began. Every week on Mondays and Fridays, an estimated 2–3,000 Haitians crossed into Dajabón to sell used clothes, cosmetics, rice, beans, and garlic, largely imported from Miami, and to buy basic foodstuffs (Dilla & de Jesus 2005). The petty trade in used clothes and shoes was mostly between Haitian and Dominican women traders. Dominican agricultural and agro-industrial producers from the Cibao valley, mostly men, sold foodstuffs, some proportion of which could not be sold in the Dominican Republic due to poor quality, type of good (e.g., rice tips), or increasing competition from US imports (Dilla 2004). As Dilla argues, the character of the trade as formal, informal, legal, or illegal "dissolved in a series of social practices of survival" for both sides, albeit the burden of need fell heavier upon Haitian traders (2004: 34). Several non-governmental organizations (NGOs) organized around the myriad of abuses – from physical harassment to arbitrary seizures of goods and taxes – that many Haitian and some Dominican traders continued to face at the hands of Dominican officials. In addition to the biweekly market, wholesale commerce in consumer and non-consumer goods such as cement, ice, rebar, and flour took place on a daily basis.

While the fraught integration of the two towns into regional circuits of subsistence and accumulation generated significant, albeit unevenly distributed, infrastructural and service improvements in Dajabón, following the embargo, Ouanaminthe's existing infrastructure buckled under the pressures of migration and commercial growth, hampered by the repeated collapse of the Haitian state. The two diesel-powered generators that provided limited electricity stopped functioning completely in 1998 and the water service had long been limited to one day a week for a fraction of the population. Despite the collapsed infrastructure, the town's commercial character was unmistakable: a host of money changers were available seven days a weeks; and three banks and five microcredit institutions were concentrated in the town center, all established since the embargo. Many homes doubled as warehouses or hotels, the town's population increased by thousands on market days, and the daily rumblings of flatbed trucks and transport trailers along the main route to and from Dajabón only ceased when border disputes led to the closure of the crossing. By the time Dominican Textile was considering the border between these two towns as a site for a trade zone, Ouanaminthe's urban population had grown approximately four-fold over a short decade, reaching between 50,000 and 70,000 inhabitants.[13]

## Reordering Trade Zones between the Dominican Republic and Haiti

In the early 2000s, DT attempted to inscribe its vision of a border trade zone onto an oxbow of the Massacre River. Local politics related to social transformations in Ouanaminthe and growing opposition to President Aristide in Haiti transformed the project into a site of national and international campaigning, causing disparate groups with conflicting interests to organize against the new development. In this section, I tease out some of these politics in an attempt to explain the circumstances that placed the trade zone at the center of both anti-neoliberal mobilizations and right-wing opposition to the Aristide government. While other scholars have discussed the politics of Aristide's overthrow in 2004 in great detail, here I consider the local context and position of Ouanaminthe as one site through which these fraught politics played out.[14]

Although generally associated with the poor *cités* of Port-au-Prince, the Lavalas movement found considerable support in Ouanaminthe in 1986 when it first emerged as a popular movement for social change inspired by liberation theology. In the volatile political opening following the flight of Jean-Claude Duvalier, a youth movement in the town formed to support Lavalas' broadly anti-imperialist and anti-neoliberal stance, the leaders of which remained active in town politics over the next two decades.[15] One former youth leader and member of an established Ouanaminthe family, Gaston Etienne, was among the many leftist, anti-imperialist students who risked their lives to support Aristide during the period. Gaston had gone to Santiago to study agronomy because positions at universities under Duvalier were only accessible through clientilist connections that he lacked. At the height of his activism, he would travel from Santiago and cross the border at night to paste pro-Aristide posters around town in Ounaminthe, most intrepidly postering the military's headquarters, a bastion of Duvalierism. As the political situation deteriorated following the military coup in 1991, Gaston remained in exile in the Dominican Republic, fearing for his life if he returned to Ouanaminthe, returning to his hometown shortly after President Aristide's reinstatement in 1994.

Aristide returned to finish the fifth year of his interrupted term supported by several thousand marines and US President Bill Clinton. Upon reassuming power, Aristide agreed to the privatization of state enterprises and to a deep cut to protective tariffs on rice as a condition of US support for his reinstatement. For these and other policy decisions, he was criticized by some observers on the left in Haiti and in the large Haitian diaspora for acceding to US demands for market-led reforms (see note 14). In Ouanaminthe, Aristide's newly formed Fanmi Lavalas (FL) party continued to win local elections. Nonetheless, middle-class and elite interests began to align to oppose Aristide during the second half of the 1990s, forming the Convergence Démocratique (CD), a coalition that spearheaded allegations of election fraud in 2000. The allegations constituted the pretext for a crippling three-year suspension of international aid to Haiti, led by an emboldened neoconservative administration in the United States and supported by the European Union and the international financial institutions (IFIs), including the World Bank and the IMF.[16]

Cut off from aid, facing pressures to create jobs, and perhaps wanting to demonstrate a willingness to support foreign investment, Aristide found an unlikely ally in the maverick Dominican investor from

Dominican Textile and his loyal backer, the powerful brand Levi's. In 2001, the Haitian government agreed to expropriate 500,000 square meters of land for the construction of the company's trade zone. The land was situated in the fertile Maribahoux plain, on an oxbow of the river across from the rapidly growing northern edge of Ouanaminthe.[17] Shortly thereafter, in 2002, the Haitian government passed a trade zone promotion law to accommodate the new investment. A local representative of the FL party, and the Haitian consul in Dajabón at the time, explained the party's motivation for the project as follows:

> All Third World countries have passed through the stage of the trade zone. Like those countries, it is a means to generate employment ... People thought that 20,000 workers would give political support to Aristide so the movement against the zone was started by his opponents, led by Guy Philippe. This mixed with popular movements who started to do paid work [*tarea pagada*] ... But those from the popular movements [i.e., Lavalas' base] understand that [the trade zone] is a necessary evil [*mal necesario*].[18]

The representative's narrative of the trade zone forming part of an inevitable process of development reflected what anti-neoliberal critics of Aristide argued to be the latter's increasingly conciliatory stance towards local and transnational capital. While the project mobilized those ideologically opposed to a trade zone model of development, some actors in the movement against the zone were indeed far more concerned that the zone would strengthen Aristide's base in the northeast, as the representative suggests.

Local opposition in Ouanaminthe was spearheaded by former Aristide supporter, Gaston Etienne, together with affected landowners, tenants, and other members of established Ouanaminthe families who formed the Komitè Defans Pitobè, or the Pitobert defense committee, named after the agricultural area of the Maribahoux plain that was expropriated for the zone's construction.[19] The plain was one of the few fertile areas of agricultural land in the mostly semi-arid northeast province.[20] Haitian agricultural production in the region was almost entirely rain-fed, meaning that the river and its flood plain concentrated the most viable areas of food production for local subsistence and commercial trade in the region. Many of the families who worked the farms around Ouanaminthe were tenants while the landowners ran businesses in town or in other cities. To make their case, the Pitobert group displayed the agricultural products of the region in protests locally and in Port-au-Prince in order to counter claims by the zone's supporters that the area was neither fertile nor under production. The committee also criticized the proposal on nationalist grounds, citing a constitutional restriction on foreign ownership of land along the border. For the Pitobert committee, as well as leaders within the local Catholic Church that had reluctantly joined the opposition to the zone, the appropriation of land was another fissure – real and symbolic – in the town's already precarious system of food provisioning.[21] More and more of Ouanaminthe's food was being supplied through the Dajabón market, a supply that could easily be cut off, and often was, due to disputes between trading interests on either side. The committee opposed the trade zone project as a development priority and called instead for the fulfilment of the long-held promise of irrigation and infrastructure to support agricultural production.

The persistent social divisions between newer migrants and long-established town residents weakened the grassroots efforts, however. Organizers of the Pitobert Committee felt that the symbolic and material importance of the Maribahoux did not resonate with newer immigrants, who, indeed, were more likely to be swayed in favor of the trade zone project by the promise of jobs. As Gaston Etienne, former Aristide supporter-turned-vocal-opponent, explained:

> The majority of people who come from outside of Ouanaminthe, if you go out and try to get their support against corruption, for example, you won't get their support even with a good campaign ... they'll say 'it's just the Ouanaminthe group that's doing this' ... This majority was ... in favor [of the trade zone], the people from outside. Because, for me, that area [Maribahoux] has a tradition, a history. I knew when there was a rice harvest, and we used to eat the crops. This was the granary for the town. For those people [from outside], this doesn't matter. They don't know the tradition.

These claims to the Maribahoux were made on the basis of the symbolic value of the town's agricultural periphery, a value that separated those 'moun Wanament' from the landless migrants who lived in the growing cités.

A separate small group of prominent local intellectuals – teachers, lawyers, and a school principal – distanced themselves from the more activist committee, claiming no qualms with a trade zone. The group believed that jobs were desperately needed in the area, but rejected the proposed location of the zone in fertile agricultural land. This group attempted to build support for the zone's relocation to Morne Casse, on the abandoned site of the Dauphin sisal plantation, approximately 10 miles west of the border towards Fort Liberté. This tract would later be the site proposed by the Haitian government as compensation to the landowners and tenants displaced by the trade zone.

Anti-Aristide forces that had united under the banner of the CD bolstered this local opposition to the zone. Two anti-neoliberal NGOs, one organized by a well-known economist who famously broke with Aristide in 1995, supplied information, raised awareness about the border struggle in Port-au-Prince, and sponsored a conference attended by 150 Caribbean activists in Ouanaminthe to support the movement against the zone.[22] Another key supporter was the staunch anti-Aristide journalist, politician, and owner of one of the largest radio stations in the northeast, Jean-Robert Lalanne. Lalanne gave the struggle ample airtime and later played a pivotal role in supporting mass strikes in Gonaïves and Cap Haïtien in late 2003 that further destabilized the Aristide government. But Lalanne, and the interests he represented, was far from the social base and concerns of many of the local activists in Ouanaminthe. Recalling Lalanne, a former Pitobert committee activist argued that their struggle had been used strategically to build the opposition to Aristide, not to support local demands.

Finally, the movement was strengthened by an independent labor union called Batay Ouvriye, translated as Workers' Struggle. Two organizers moved to Ouanaminthe, one from Cap-Haïtien and one from Port-au-Prince, to support the campaign against expropriation and to build a base for eventual trade union organizing inside the trade zone. One of the organizers, Yannick Etienne, was from an old Ouanaminthe family, exiled under the first Duvalier regime.[23] Etienne and other Batay militants had spearheaded organizing campaigns in the trade zones in Port-au-Prince in the 1980s. Initially supportive of the Lavalas' movement, Batay became one of the most outspoken left critics of the Aristide administration. In sum, the trade zone project became a sort of touchstone for a multitude of grievances that united groups of disparate political orientations – anti-Aristide, anti-neoliberal, nationalist, and localist – in their opposition to the project.[24]

Although Dominican Textile was able to break ground in March of 2003, the project had helped to galvanize opposition to the very administration that had paved the company's way in Haiti. While the movement against the zone failed to stop its construction, the political upheavals in Ouanaminthe and Haiti in 2004 would impede the project's operations. Two important consequences of these early mobilizations against the zone affected its subsequent operations: first, several of the students who had organized with the Pitobert committee were among the first hired at the trade zone when it began operations at the end of 2003. Batay Ouvriye worked with this group, informally referred to as the zone's "first class," to form a union committee. Second, the company alienated many notables in town as well as the Catholic Church, both in Dajabón and Ouanaminthe. For political support to resolve the many conflicts that lay ahead, the company would have to turn to distant allies in Port-au-Prince and Santo Domingo.

## CODEVI: Haitian State Collapse and Unionization

By the end of 2003, the basic infrastructure for the new trade zone, named the Industrial Development Company (CODEVI) was in place. The bi-national zone gave expression to a strict territorial division of labor. In addition to the 400,000 square meters of land on the river's oxbow on the Haitian side of the border, where factories were to be located, the company acquired an adjacent 150,000 square meters on the Dominican side, thereby both guaranteeing bi-national access to the zone and privatizing a small section of the political boundary.[25] Haitian workers and a handful of lower-level managers entered and exited from Haitian territory, crossing a bridge over the river and passing through a small enclosed area within the zone that housed the training center, a clinic, and a corporate office (and later, the radio and TV stations) (see Figure 5.1). Dominican supervisors and managers, and two higher-level Haitian managers, entered and exited from the Dominican side. Haitian and Dominican customs houses faced each other near the Dominican entrance, signaling the territorial boundary between the two nation-states. The first two factories were installed by the end of the year. One plant had an exclusive contract with US-based Levi's to produce jeans and the other was established jointly with US-based Hanesbrand company to produce t-shirts.

**Figure 5.1**  Haitians exiting the CODEVI trade zone. The siting of the zone on the river's oxbow privatizes a section of the border, allowing the administration to control Haitians' access to the zone through a privately built and guarded bridge. Source: Photo credit: Jude Stanley Roy. Reproduced by permission of Ayiti Kale Je/Haiti Grassroots Watch.

By this time, Dominican Textile was close to exhausting its capital resources a result of its investment in the trade zone project as well as the installation of more capital-intensive processes in Santiago, including a cutting room transferred from Miami, a small knit mill moved from Alabama, and a product development department (see Chapter 3). The company was having difficulty servicing its debt and turned to the World Bank's private-sector lending arm, the International Finance Corporation (IFC), for a loan. The IFC underwrote a 41.4 million dollar financing package, including a 20 million dollar low-interest loan, to continue upgrading the company's facilities in the Dominican Republic, refinance its debt, and complete the construction of CODEVI. In addition to the IFC loan, the bulk of the other monies were to be generated during the first two years of the zone's operation (IFC 2003a). The loan projected the creation of 1,500 jobs in the initial phase of the trade zone and the possibility of catalyzing investment to generate 20,000 direct jobs once the entire site was developed. In IFC project documents, border zones were suggested as possible palliatives for the migration of Haitian workers to the Dominican Republic. The IFC also claimed that its involvement would help to manage social and environmental risks, arguing that its "extensive environmental and social due diligence and sustainability undertakings provide comfort to [Dominican Textile's] key customers helping confirm the [company's] adherence to best industry practice" (2003a: 4).[26] The IFC (2003b) justified the loan as a contribution to the stability of the Dominican garment export sector by "securing the long term growth" of its largest firm in the face of shifts in the industry. The project's authors wrote that "[b]y providing long term corporate funding IFC will bolster the financial sustainability of Dominican Textile, the largest employer in the Dominican Republic" (IFC 2003a: 4).

The irony of a loan for a company to build its production capacity in Haiti justified by its importance as the largest employer in the Dominican Republic was not lost on Dominican unionists who had undertaken several failed attempts to organize in Dominican Textile's 14 plants in and around Santiago. Following the signing of the loan agreement, the Dominican Federation of Trade Zone Workers (FEDOTRAZONAS), which had spearheaded organizing efforts in Dominican Textile's Santiago-based operations, began to criticize the loan agreement, arguing that the company did not respect freedom of association.[27] The union federation and its international allies, including the AFL-CIO and the International Trade Union Confederation (ITUC), pressured the head of the IFC to respond to the Dominican union's allegations. The campaign resulted in an independent investigation of the allegations and, once verified, a public statement from IFC that the loan would subsequently be conditioned on respect for core labor standards by the beginning of 2004 (IFC 2005; see also IFC 2006).[28] In the revised loan agreement, Dominican Textile agreed to remediation, including training for managers and supervisors on labor rights and regular audits.

By the end of 2003, Haitian workers, supported by Batay Ouvriye, formed a committee and began to agitate around basic labor demands: wages and hours, sexual discrimination and harassment, and the treatment of Haitian workers by their primarily Dominican supervisors. The workers' committee also objected to the presence of the Haitian Ministry of Social Affairs and Labor in an office inside the privately owned trade zone, demanding a formal territorial separation between the state and the company.[29] While frequent tensions between Haitian workers and Dominican supervisors galvanized the organizing effort, as did the support from community leaders who had opposed the zone, in the early formation of the union, the campaign was couched largely in terms of labor-management conflict and basic labor rights. The union gained support among workers quickly and submitted its paperwork to the Labor Ministry for recognition as the Union of CODEVI Workers (SOKOWA) at the beginning of February 2004.

The possibility of resolution through the workings of the civil state dissipated with the Haitian state's collapse, however. Confrontations between pro- and anti-Aristide forces escalated throughout the country in early 2004. In Ouanaminthe, these confrontations materialized in violent attacks against opponents of the trade zone by pro-Aristide forces, including the burning of Gaston's pharmacy in early February. One week later, a rebel militia entered Ouanaminthe from the Dominican Republic, led by Guy Philippe and other former members of the military (see note 18). Several prominent members of the town and at least one well-known activist from the Pitobert committee joined them. The rebel forces proceeded to CODEVI where they found themselves facing the Dominican military, called in to defend the trade zone, in a standoff on the bridge (Mañaná 2004). After tense negotiations, the rebels received fuel and other supplies from the trade zone management and retreated.[30] Following this confrontation, the Dominican military began to have an active presence inside the trade zone, technically on Haitian soil, quickly transforming the union campaign into a struggle over national sovereignty. What ensued in this climate of instability is contested by all sides, but the initial effect was the firing of 31 union leaders in a militarized and violent conflict.

Pressure from various US-based labor rights groups and unions, together with the IFC and Levi's, the continued organizing campaign in Ouanaminthe, and mediation by local leaders led to the eventual reinstatement of the fired workers and the apparent demilitarization of the border zone.[31] The union began to have preliminary meetings with management, the latter acceding to some of the workers' demands.[32] The union claimed that the company still refused to negotiate over their substantive grievances, however, including wages and the incorporation of reinstated workers into the main operations of the plant.[33] A team of three independent local monitors maintained a permanent presence in the factory for five weeks, holding weekly conference calls with Levi's, Dominican Textile, the IFC, and US-based labor rights groups. The situation escalated, however, as allegations of violent treatment between Dominican managers and Haitian workers, especially women, circulated in union leaflets, and the conflict appeared to spread to the second plant producing for Hanes.

In early June, workers staged short work stoppages, alleging the company had refused to address their immediate grievances by cancelling a meeting with the union. The union claimed that the Dominican military forcefully expelled four supporters, all women, who were participating in the job action. In protest, the union called for a strike the following Monday. A temporary solution was mediated by the Haitian consul and workers agreed to return to work. On June 8, Dominican Textile sent out a public letter announcing the closure of the trade zone, citing "repeated threats and violent action that threaten employees on the part of a radical activist group called Batay Ouvriye" (on file with author; also Santana 2004). The company also publicized the visit of the archbishop of Santo Domingo (*Hoy* 2004), who gave the project his blessing in an attempt to minimize the effect of the local Catholic Church's opposition. Three days later, half of the 500 workers at the Levi's plant were fired while Hanes announced it would discontinue orders.

Numerous meetings between different actors ensued in what had become a highly publicized, international conflict. The IFC, while reluctant to cancel the loan, had also publicly conditioned the loan on the respect of freedom of association during the previous campaign started by Dominican unions. By February of the following year, Dominican Textile agreed to reinstate 150 of the workers fired the previous June, and began the process of negotiating a collective bargaining agreement with the union, signed in December of 2005. The three-year agreement guaranteed a base wage of 900 gourdes (23.68 dollars) per week and annual increases that would offset inflation, in addition to union recognition, a labor-management conflict resolution committee, and protections against sexual harassment.[34] The collective agreement was the only one of its kind in the Haitian garment sector, and among Dominican Textile's factories in the Dominican Republic and Haiti.

## Disposability at the Border: Managing the Trade Zone[35]

In late 2006, after several months of negotiation with Dominican Textile, I was permitted limited access to the border trade zone. My conversations were confined to Spanish speakers because of my language ability and my presence was greeted with considerable suspicion stemming from the highly publicized international labor rights campaign two years prior. Production in the trade zone followed similar outsourcing patterns that had characterized the industry since the beginning of the export apparel boom in the 1980s with the exception that knit fabric was now partly being milled in the Dominican Republic. In the t-shirt factory, knit material arrived from a Hanesbrand mill in the United States (via the Dominican Republic) and from the Dominican town of Bonao where the US manufacturer had established a 400-person facility as part of its "co-production" strategy with Haiti. Receivers weighed the cut material and logged the shipment into the North Carolina-based company's tracking system, triggering a 10-day put-through clock. The factory was not yet at full capacity and dozens of sewing machines from Hanesbrand, marked with shipping labels from a previous journey from Texas to San Pedro, Mexico, stood crated in one corner awaiting the expansion of production. In the jeans factory across the way, stacks of pieces of Levi's 501s and 505s were being logged in at the far entrance to the plant. The material was all imported from the United States, since the Dominican Republic had not been able to attract a woven mill – a major, capital-intensive investment – to its shores (see Chapter 3). Workers in Dominican Textile's plants just outside Santiago had cut the material, mounted the leather labels, and, for the 505s, assembled the back pockets with a special machine. The finished jeans and t-shirts, stacked high on pallets, would be trucked to the Dominican Republic's largest port near Santo Domingo and transported by ship to the United States.

The gendered, national, and racialized division of labor that pervaded the logic of the zone operated as a micropolitics at the level of the factory floor. Dominicans, mostly men except for a handful of women in quality control, worked as supervisors, auditors, dispatchers, and in maintenance and inventory. All the operators were Haitian. Eighty percent of the operators making t-shirts were women while men made up 40 percent of the operators producing jeans. I was told that the difference in the gender ratio between the two factories had to do with the quality of material – knit material for t-shirts is lighter than the woven material of jeans. This logic, which operated in the Dominican Republic as well (see Chapter 3), normalized a gender wage gap: operators in the t-shirt factory on average earned about 30 percent less than their counterparts in jeans. The latter benefitted not only from higher piece rates, but also more specialized, and therefore relatively better paid, operations.

One expression of the micropolitics that underpinned the hierarchical division of labor of the zone was the set of actions surrounding drinking water. Each of the areas on the factory floor where Dominicans worked had bottled water stations covered by a cardboard box or a cloth. One of my guides, an assistant in human resources, offered me an unsolicited explanation: while there was nothing wrong with the water consumed by Haitian workers through a ground filtration system, Haitian water could harm Dominicans who were accustomed to "their" water. The efforts to hide the water stations suggested otherwise, and several weeks after my visit, a significant conflict arose when Haitian workers blockaded the Aguas Beler water delivery truck from Dajabón following several days where the water system used by workers in the Levi's factory had been inoperable.

Perhaps it was conflicts like those over water – conflicts that were inseparable from the national difference between management, on the one hand, and the workforce, on the other – that prompted Dominican supervisors and managers to emphasize the temporary nature of their work there. I was told repeatedly that the goal of Dominican Textile was to make the plants Haitian-run as soon as possible. No managers expressed any reservations about this projected change, one that would, given the company's trend towards shutting down factories in the Dominican Republic, likely leave them without jobs in the sector. I soon realized that very few members of management and supervision had been working for more than a year in the trade zone; the general managers of the Levi's and Hanes factories had been replaced two and three times respectively, and I met no Dominican supervisors who had been at the zone more than 18 months. The high management turnover suggested that work in the border trade zone was either not highly desired by the company's Santiago-based employees, or that one strategy to deal with conflicts was to replace managers and supervisors, or a combination of both.

One exception to the revolving door of management was Sergio, the company's head of human resources for the zone. Sergio was the first Dominican Textile employee to arrive in Ouanaminthe before construction was completed; he was charged with hiring and training Haitian workers for garment assembly. Sergio had set up four training centers in Santiago in the previous decade during the company's expansion there. Like many of his male counterparts whom I met in the company, he had started as an operator in his late teens and worked his way up the highly gendered ladder of management.

Sergio was the only Dominican manager who had spent time in Ouanaminthe, primarily during the initial phase of the project when he had set up a provisional training center at one of the town's two night clubs, and who still visited on occasion. His views of the project and the town echoed those the company president had expressed to me in Santiago: "Go to Ouanaminthe on any day of the week," he told me, "and it looks like a Sunday afternoon. Everyone is just hanging around, not doing anything." Sergio thus saw himself as contributing to the much needed development of the town and its inhabitants by providing training and jobs to an otherwise apparently unoccupied workforce. Escorting me between the Levi's and Hanes plants one day, Sergio was handed a request for a new operator from a Haitian supervisor that he had trained as an operator three years earlier. "Look," he showed me the form, "now [the supervisor] has good judgment [*tiene criterio*]. He wants someone who is punctual." Sergio felt his efforts were facilitating the natural maturation of the workforce from a state of indiscipline and non-productivity to one of responsibility and efficiency. He expressed deep satisfaction in how the factories were coming up to speed and how workers who once "couldn't comprehend producing hundreds of pieces in a day" were producing just as well as their Santiago counterparts.

Standing on the bridge that crossed the Massacre River at the Haitian exit of the zone just before the end of the shift one day, Sergio pointed to the women, dressed from the waist down, who were washing clothes in the river below. "When I first arrived, there were many more [washer women]," he told me. "I talked to all the different [church] denominations [in Ouanaminthe] and asked them to create awareness with the people so that they wouldn't wash clothes like that near the zone. It creates a bad image." He continued, "How can we be sewing fashionable clothes here next to so many naked [sic] people? ... For foreigners and Dominicans, it's very shocking." Sergio seemed to see no contradictions between what he perceived as "idle" labor in Ouanaminthe and his own efforts to influence where and how women forged their livelihoods through daily, often paid, tasks of social reproduction. Work in Ouanaminthe, revolving around these activities and the multiple forms of livelihood that ebbed and flowed with exchange rates, prices, and politics of the bi-national market, was conveniently reduced to a latent source of labor power in management's view of the town.

Like Sergio, Martin, one of two Haitian human resource managers, seemed to perceive his role as facilitating worker development, but expressed reservations about the latter's prospects through garment employment. Martin was originally from Port-au-Prince and lived in Santiago where he had gone to university. He had been working at the trade zone for one year. Martin lamented the low level of education of the zone's current employees who mostly came from rural areas. One of his hopes was for a second shift to be established in the zone because, as he expressed to me, "nobody in Ouanaminthe was going to sacrifice their schooling to work in the trade zone." Martin shared his own desires to continue his studies, hampered by the long hours demanded in his current job. Several interviewees in Ouanaminthe echoed Martin's perception that the trade zone offered limited opportunities for the upward mobility of Haitians.[36]

Martin's office received a steady stream of workers, mostly accompanied by supervisors, asking for permission for special leaves or severance pay due to illness or injury. The tension between extracting surplus value from labor and the need for workers to be socially reproduced as able bodies played out repeatedly in these daily, mundane interactions. One morning while I was in his office, a woman came to see Martin without her supervisor and showed him her swollen foot. She asked for permission to come to work a half-hour late each day in order to seek treatment until the foot healed. Martin agreed. A few minutes later the injured workers' Dominican supervisor came into the office and demanded that Martin reverse his decision: "Who knows when [the foot] will heal?" he complained. Martin replied curtly that the worker would produce better with a healthy foot and the supervisor retreated to find a more senior manager in the hopes of overriding Martin's decision.

Representatives from Santiago arrived at the trade zone weekly to solve problems and to deal with continued conflicts between workers and management. The delegation of Santiago managers that I accompanied on one of their weekly trips to the zone was led by the company's division head, José Antonio. When he arrived in his white SUV at the factory, José Antonio was greeted as "el ministro" by zone managers and union leaders. He laughed at the designation, which signified the importance of his role as an arbitrator and decision-maker. He immediately faced an all-male union delegation demanding an audience with him to discuss recent conflicts. Later on, he shared that the biggest problem the company was dealing with in the zone was frequent work stoppages. "[The workers] stop easily," he commented, "but we are getting a handle on it." Details were not made available to me, but clearly the practices of negotiated accommodation and training, represented in my interactions with Sergio and Martin, were not the only forms of power exercised to keep production going. Several strategies were evident. First, the company had set up a TV and radio station inside the trade zone. The radio station allowed management to control the audio content on the factory floor and, together with the TV station, to extend the company's influence in the northeast region of Haiti where the trade zone project continued to be controversial. Second, the company continued its practice (as in 2004) of mass firings as a strategy to deal with conflict. In a subsequent interview with the Haitian Ministry of Social Affairs, I learned that the company had officially reported firing 170 workers in the first four months of 2007, nearly 15 percent of the workforce at that time. The bulk of these workers were fired for carrying out a work stoppage at the beginning of the year when the company attempted to change how the daily production totals would be tallied. The revolving door of workers was evident in the streets of Ouanaminthe. Grey trade zone t-shirts, given to each worker when they began to work at the zone (and deducted from their salary if they were lost), were worn by the hundreds in town on a daily basis, markers of the pace at which the company fired workers, and disgruntled workers sought out livelihoods in commerce and other activities.

## Conclusion: Reworking the Geographies of Coloniality

Historical and contemporary geographies of violence, the force of the embargo, structural adjustment in Haiti, and the fractiousness of the Haitian state created the conditions of possibility for this "new" low-wage frontier in Ouanaminthe. These factors were linked contingently to processes to renew accumulation in the Dominican Republic spearheaded by the state and export capital. These efforts included the production of discourses and policies that rendered feminized garment labor as superfluous both to the *mentefactura* paradigm advanced by the Dominican state and the World Bank and to the restructuring process in the Dominican garment sector. In short, the trade zone was far from a naturalized expansion of wage labor to a new frontier. Rather, garment capital, supported by its transnational counterpart and international development donors, seized upon the border – an intensifying contour of sociospatial unevenness – as a source of potential for the extraction of surplus value.

In his formulation of the notion of the coloniality of power, sociologist Anibal Quijano argues that the patterned nature of such efforts is inseparable from the *longue durée* of colonialism. Quijano (2000a) argues that the raced and gendered hierarchies forged through the conquest of the Americas constitute a terrain of articulation, one that adapts historical patterns of exploitation and domination to the necessities and attendant conflicts of contemporary capitalist accumulation. Elsewhere, he elaborates:

> The distribution of social identities [through racial categories] would henceforth sustain all social classification of the population in America. With and through it, diverse forms of exploitation, labor control and relations of gender would be articulated in changing forms depending on the necessities of power in each period. (Quijano 1998: 30)

Yet the precise workings, forces, and limits that constitute this abstract terrain of articulation described by Quijano can only be understood by examining actually existing geographies of uneven development (see Restrepo & Rojas 2010). As we have seen in Ouanaminthe, the efforts of capital and the state to arrange places and socially marked bodies into a hierarchical ordering of value were inseparable from the practices of resistance and livelihood that surrounded and exceeded these efforts.

The shifting contours between hyper-exploited wage work and the exclusion of people and places from this relation – contours I have traced between Santiago and Ouanaminthe and in the daily operations of the border trade zone – were shaped by multiple kinds of social relations. These included class struggles, congealed through intersecting gendered, racialized, and national hierarchies, as well as state violence and foreign intervention. In the case of the border trade zone, the very conditions of possibility for its establishment – that is, the embargo, the reproduction of Ouanaminthe as a pole of migration and as a regional market, and the fractious cross-border politics that accompanied this process – continued to shape, and indeed to limit, the fixing of accumulation to this margin of intense unevenness. These counter-tendencies manifested themselves both at the level of organized protests, like those of the Pitobert committee and the garment union, and in the everyday practices of managers and workers. My account identified some of these latter, everyday limitations to the project of accumulation at the border: the continual replacement of management, the waves of firing of workers, the barriers that workers faced to reproduce themselves as able, laboring bodies in the context of Ouanaminthe's collapsed infrastructure, and the unmet expectations of Haitians for social mobility through work in the trade zone.

The expansion of the global garment production network to the northern border region provides one example of how coloniality is materialized spatially in unexpected ways. We must not forget that only a decade before the first crops were uprooted in Pitobert to make way for industrial investment, the idea of such an arrangement of labor and capital was unthinkable. Not only was the relatively isolated border town of Ouanaminthe an inadequate source of labor, but Dominican and transnational capital was supported by the Dominican and the US states to exploit Dominican labor within the garment production network.

Indeed, seen from the perspective of Haiti following the 2010 earthquake, the border trade zone, and its troubles, was a sort of bellwether for what was to come. With the devastation of Port-au-Prince by the earthquake, the country's northern region was being reproduced as an ideal site for export production. With large aid flows directed towards Haiti, the size and scope of the new investment projects – promoted under the banner of "reconstruction" and subsidized by IFIs and the US government – dwarfed the efforts of Dominican capital in Ouanaminthe. And while not enough time has passed to understand how these designs for development in the region will intersect with the area's layered histories and transforming social relations, the intense focus on export assembly factories as a form of reconstruction in Haiti is discouraging. Analysis of developments after the earthquake returns us to concerns of global production as a discourse of transition rather than connection, a dilemma to which I will return in the following chapter.

## Notes

1   This chapter contains materials previously published in Werner (2011).
2   The export goods sector based exclusively in Port-au-Prince – producing not only garments, but also toys and electronics – thrived in the 1970s and early 1980s, employing some 40,000 workers, 40 percent of whom worked in garment assembly (DeWind & Kinley 1988). For more on this earlier period of export production, see Chapter 2. For the legacy of this period in the discourses of reconstruction following Haiti's 2010 earthquake, see Chapter 6.
3   With the exception of the trade zone discussed in this chapter, all other investment attempts were made in the Haitian capital of Port-au-Prince, the site of the country's beleaguered, but long established, export assembly sector. Among the largest investments were Hanesbrand (formerly Sara Lee), a US-based manufacturer, and Gildan, a Quebec-based manufacturer. Both companies built large knit mills in the Dominican Republic and shipped part or all of the material spun in their knit mills to wholly or jointly owned facilities in Port-au-Prince. Dominican Textile's two main Dominican competitors, DCD and IA Manufacturing, both attempted, unsuccessfully, to set up joint ventures or stable subcontracting relationships in Port-au-Prince. Decentralizing the sector and focusing new investment on the north of the country only became a major strategy of transnational capital following the 2010 earthquake (see Chapter 6).
4   The arrangement of labor and materials both at the border and in the new facilities in Port-au-Prince was facilitated by the Dominican Republic and Central America Free Trade Agreement with the United States (DR-CAFTA), which the Dominican garment sector had a key role in negotiating (see Chapter 2). The agreement included so-called co-production with Haiti, or garment assembly in Haiti of textiles otherwise transformed or shipped from the Dominican Republic, still made largely of imported US fabric or yarn.
5   I am greatly indebted to the research assistance of Mr Delien Blaise in Ouanaminthe, to whom this chapter owes its emphasis on the transformation of Ouanaminthe's social structure after the embargo. For more on methodological issues, see my discussion in Chapter 1.
6   See Chapter 2 for the key literature on the massacre, including the role it played in consolidating Dominican peasant identity and the Cibao as a locus of *dominicanidad* (Dominican-ness).

7   For an ethnographic account of Dominican border migrants in the Santiago trade zone, see Derby and Werner (2013).

8   This is a statement based on Ouanaminthe residents' accounts. According to Saintilus (2007), the Haitian authorities did not officially close the border until 1963 and it was re-opened when Jean-Claude Duvalier took power following his father's death in 1971. In either case, both interviewees and Saintilus describe transit across the border as greatly impeded, controlled exclusively by the Haitian and Dominican militaries.

9   There was no consistent account of what Haitians primarily bought in the Dominican Republic at the beginning of the border trade. Also, some interviewees described the market during the period between Duvalier's departure and the embargo (1986–1991) as largely contraband, controlled by both militaries. According to Saintilus (2007), the bi-national market days were established in 1988.

10  The statistic is from Silié and Segura (2002: 73). Interviews with Dominican exporters to Haiti suggest that tariffs were relatively arbitrary. The reduction of tariffs on rice imports to Haiti during the post-dictatorship period spurred a vigorous rice trade with the Dominican lower classes, which intensified after the second round of tariff liberalization following the embargo. At the time of my research in the mid-2000s, the use of "Miami" rice transshipped through the border was widespread among food vendors in the informal sector around the trade zone in Santiago.

11  The United Nations joined the embargo in 1993. Until that time, while there was a gasoline trade through Dajabón and Ouanaminthe, there were still tankers arriving from non-OAS member states. The UN embargo lasted from June 1993 to October 1994 although it was suspended for two months from August to October 1993. Balaguer, while officially supporting the OAS resolution, also sanctioned the passage of food and gasoline through Dajabón. Once the embargo was tightened, Balaguer's policy remained in place, albeit unofficially (Sagas 1994). The *New York Times* reported that the gasoline trade happened in full view of the Dominican military (French 1994). See Chapter 2 for background on Balaguer's anti-Haitianism during the period of Trujillo's rule.

12  By the summer of 2009, the company president had launched his candidacy for the region's senate seat affiliated with the ruling Partido de la Liberación Dominicana (PLD).

13  The *commune* of Ouanaminthe has five rural *sections communales* plus the town. The 1982 census registered 36,495 inhabitants in the commune, the majority of which would have been rural inhabitants. The next census in 2002 (published in 2005) registered 82,549 inhabitants in the commune, with 43,774 in the town. The numbers here are based on estimates given by various informants, including two who participated in the 2002 census, taking into account the widespread claim that the census was not accurate, as well as the perception of a booming town that would lead informants to overestimate the population. The population at the time in neighboring Dajabón was 16,500.

14  For a comprehensive, fascinating account that spares neither criticism of Aristide nor ire for US and European imperialism, see Dupuy (2007).

For an incredibly detailed and documented account sympathetic to Aristide and deeply critical of left intellectual criticism of his second mandate, see Hallward (2007). For a more general journalistic overview, with particular attention paid to the role of US intervention, see Chomsky et al. (2004).

15  The movement was called *Mouvement des Jeunes Patriotes de Ouanaminthe* and was affiliated to the national federation of popular organizations.

16  For more on the aid suspension and the United States' role in efforts to topple Aristide's second government, see Chomsky et al. (2004).

17  While the Massacre River generally runs along the border, the siting of the zone on the oxbow placed the border on the zone's eastern side and the river between the zone and the town, a location that planners felt provided extra security from the town since the zone could only be accessed by a bridge controlled by the company.

18  Guy Philippe was a former police chief and military officer from Cap-Haïtien who was one of the leaders of the coup against Aristide in 2004, which he organized from exile in the Dominican Republic (Dupuy 2007: 152).

19  The number of farmers displaced by the trade zone was a matter of great dispute. The company initially claimed that 55 farmers were affected and offered compensation to the landowners only (number unknown). This total was subsequently challenged. The committee argued that not only were more landowners dispossessed but that dozens of tenants lost their livelihoods. The International Finance Corporation (IFC; see below) reviewed the compensation plan once it became involved in the project in 2003 and found that 95 farmers were affected and merited compensation. The representative from the Haitian agricultural ministry in charge of finding alternative land for owners and tenants dispossessed by the FTZ said the government's plan was assisting 172 affected families, 100 tenants, and 72 owners. For the IFC assessment, see IFC (2003a).

20  The rest of the region is similar to its Dominican homologue, the dry *línea noroeste*.

21  Father Regino Martinez, then the Jesuit leader of a large human rights NGO in Dajabón, and leading figure in social movement politics on the border, initially supported the project on the condition that it did not expropriate fertile land. Although receiving assurances to the contrary, Father Regino publicly withdrew his support when the trade zone's location was finally clarified.

22  The conference was a regional assembly of the Convergence of Movements of the Peoples of the Americas (COMPA), titled "Building Alternatives in the Face of Neoliberalism" and met in Ouanaminthe, October 14–16, 2002.

23  Batay Ouvriye is a Haitian labor organization started in Port-au-Prince. Several of the lead organizers, including Yannick Etienne, are return exiles from the United States. Etienne spent 25 years in New York City during which time she attended CUNY and participated in anti-war and anti-imperialist movements in the 1960s. She returned to Haiti in the final years of the Duvalier regime to organize against the dictatorship.

24 These local struggles over land and their contribution to de-legitimizing Aristide's government resonate with the work of Michael Levien (2013), who posits the role of the state in expropriating land for neoliberal market purposes as a specific regime of dispossession.

25 Dominican Textile successfully lobbied the Dominican government to expropriate the land from a Dominican investor in the import–export business. In addition to guaranteeing direct access to the zone from within Dominican territory, the land was used to build housing for Dominican personnel.

26 Because the IFC lends to private companies and not governments, the institution benefits from the kind of flexibility that created the possibility to finance the border trade zone while the donor community withheld aid dollars from the Aristide government. The low-interest loan was the determining factor in the viability of the company, allowing it to weather a credit crisis that consumed several of its competitors. Project documents state an interest rate of 3.5 percent per annum, relative to a commercial lending rate of between 14 and 18 percent. By 2007, Dominican Textile would emerge as the most important firm in the sector, concentrating approximately one-fifth of what was left of the Dominican Republic's garment export sector.

27 FEDOTRAZONAS was partly funded by the US labor federation, the AFL-CIO, specifically its international arm called the Solidarity Centers. Since reforms to the Centers following the end of the Cold War and a change in the AFL-CIO's leadership, the Centers have supported local union organizing in the circum-Caribbean garment sector. These efforts have not been entirely disinterested, but should also not be simply dismissed as US labor protectionism since the local unions receiving these funds are also strategic actors and not simply agents of the AFL's agenda. See Traub-Werner and Cravey (2002) for a critical analysis of Solidarity Center-funded organizing in Guatemala's export garment industry. For an analysis of the fraught politics of US labor internationalism and its legacy, see Glassman (2004).

28 The IFC's previous position is expressed in the full project report written to the World Bank's Board of Directors for the purposes of loan approval. The authors give their preliminary assessment that these claims are baseless but concede to recommending an independent investigation. "We will only commit to the proposed investment," they write, "if this investigation confirms our provisional assessment and the allegations are shown to be unfounded" (IFC 2003a: iv).

29 Union leaflet, December 4, 2003. On file with author.

30 Several local organizers said that the militia subsequently participated in violent repression of SOKOWA leaders.

31 The head of security of the zone at the time of fieldwork claimed also to be an active duty officer in the Dominican military. Author's conversation with head of security, May 7, 2007.

32 Meeting minutes between SOKOWA and CODEVI, April 27, 2004. On file with author.

33 Union leaflet, April 30, 2004. On file with author. The reinstated workers were apparently being kept in the training center in order to limit their influence on the rest of the workforce.

34 Convention Collective de Travail, Decembre 2005. On file with author. Average wages for Dominican Textile's Dominican operators in Santiago at the time were approximately 1,500 pesos per week, or about 53.57 dollars depending on production.

35 Individuals' names in this section are pseudonyms.

36 The principal of the main high school, for example, explained that fewer of his students were seeking work in the trade zone. Seeing themselves as future professionals, they felt the zone would not offer the opportunity to work and study at the same time.

# 6

# Haiti, the Global Factory and the Politics of Reconstruction

## Introduction: Shaking Ground, Breaking Ground

On January 12, 2010, at 4:53 pm, an earthquake devastated the already precarious foundations of urban development in Port-au-Prince and the surrounding region. The damage was extensive, with as many as 220,000 people killed and an estimated 600,000 internally displaced. One day before the first anniversary of the disaster, the US government, the Inter-American Development Bank (IDB), and the Interim Haiti Recovery Commission (IHRC) announced their showcase development project with the support of Haiti's newly elected government: a trade zone in the country's northern region called the Caracol industrial park, named after the adjacent bay. The project combined foreign investment from the Korean conglomerate SAE-A with IDB and US government loans for infrastructure development, creating a trade zone five times the size of the neighboring border trade zone in Ouanaminthe (see Chapter 5). Although the zone promised to generate employment in various sectors, the main activity was to be apparel assembly. SAE-A committed to creating 20,000 garment jobs, promising to make the company the country's largest private sector employer. In addition, the US government pledged 120 million dollars towards a power plant to service the zone and housing for workers, and the IDB placed the project at the center of a 100 million dollar aid package to promote the country's northern region as an economic growth pole (CIRH 2011; De Córdoba 2011).

The boosterism surrounding the project was evident in the high-profile groundbreaking ceremony later that year. The images of the Caracol

*Global Displacements: The Making of Uneven Development in the Caribbean*, First Edition.
Marion Werner.
© 2016 John Wiley & Sons, Ltd. Published 2016 by John Wiley & Sons, Ltd.

inauguration betray the weight of foreign capital backing the development. A Caterpillar front-loader towers over the all-male crowd, including President Michel Martelly, Prime Minister Gary Conille, and former US President Bill Clinton, all dressed casually in open-collared button downs and polo shirts, jeans, slacks, work boots, and hard hats. Each of the three politicians wields a shovel in turn, vigorously pushing concrete down a trough from a large mixer to fill the foundation of the ceremonial corner stone. A large billboard displaying a simulated image of the finished installations hovers above the crowd. The ceremony was widely reported in major print and television media in the United States. Haiti, we were told, was open for business.[1]

This chapter analyzes shifts in development policy after the earthquake in Haiti, which ultimately re-instantiated the global factory as a model for Haitian development. I draw upon national and international donor development plans together with media reports as well as interviews with development actors and civil society leaders. As a discourse, the global factory reduces dynamic geographies of uneven development to understandings of change as a series of successful or failed transitions along a sequence of stages. My analysis of debates on reconstruction after the earthquake reveals how this discourse operated in the context of the country's devastation. I argue that the promotion of new apparel factories as a response to the disaster came at a very high cost, that of reinforcing Haitian workers' subordinate integration into global production networks, thus hampering efforts to promote alternative models of development.

The chapter develops this argument in four sections. In the sections titled "Poverty Reduction" and "Trade Zones as Containment," I analyze the place of assembly production in Haitian development plans prior to the earthquake. "Poverty Reduction" examines the context and content of Haiti's Poverty Reduction Strategy Paper (PRSP), produced in 2007. Haiti's PRSP represented a minor shift in development policy by placing significant emphasis on domestic agricultural production and rural development. Like much so-called post-Washington Consensus policy, the document framed a development agenda substantively focused on growth together with policies directly aimed at reducing poverty. In "Trade Zones as Containment," the heart of the chapter, I turn my attention to the rise of "economic security" as a defining narrative of development in Haiti, following a devastating hurricane season and a consumption crisis in 2008. This narrative, advanced in the widely circulated Collier report (2009), placed the global factory at the center of the country's development priorities as an imperative to secure the conditions for economic growth. The Collier report dismissed the embedded liberal discourse of poverty reduction in Haiti that had been advanced in the country's PRSP, which was never implemented. This development-as-security narrative rested upon the notion of trade zones as a way to contain Haitians, thereby constructing them as a racialized threat and ultimately reproducing age-old policy patterns between Haiti and US and European powers. By mobilizing the global factory as simultaneously an economic development and a security priority, the Collier report's vision gained prominence in Washington-based donor circles following the earthquake.

Given the devastation in Port-au-Prince as a result of the earth quake, the Haitian government renewed the long-standing, but never implemented, priority to de-concentrate economic activity, as part of a broader mandate of decentralization. In the section "From Poverty Reduction to Poverty Redistribution?," I place this priority in its particular context in Haiti, and explore how it was taken up by donors, and in particular, the IDB, the largest donor in the field of trade promotion and private sector development. I argue that donors mobilized the notion of decentralization in this context as a kind of shell game, marginalizing policies that would decrease poverty and extend services to rural areas while prioritizing "spatial equity" in investment. In the conclusion, I return to my argument about the global factory as a development discourse shaping contemporary debates and policies on Haitian reconstruction. Trade zone promotion in Haiti rests upon the assumptions of Eurocentric development and the priorities of multinational capital. This global factory discourse positions Haiti in the waiting room of "progress," perpetuating the fiction that the country's troubles are due to its exclusion from capitalism, rather than a product of Haiti's subordinate integration into geographies of uneven development. While the country's position is certainly not fixed, the policies of post-earthquake reconstruction – anchored to the global factory – contributed to constraining the political space to challenge this arrangement.

## Poverty Reduction: An Embedded Liberal Moment

By the late 1990s, the much-vaunted Washington Consensus on market-led regulation was facing a legitimacy crisis, prompted by oppositional social movements and poor economic performance in much of the global South. As a result, the World Bank and the IMF broadened their objectives from growth and macroeconomic stabilization to "poverty reduction," achieved by more active governance of market institutions.[2] In addition, the IFIs mainstreamed participatory approaches following critiques of top-down policymaking (Porter & Craig 2004; Craig & Porter 2006). In line with their new approach, the IMF and the World Bank implemented Poverty Reduction Strategy Papers (PRSPs): development plans to be written by host country governments based upon broad consultation with civil society groups, but nonetheless guided by, and ultimately requiring the approval of, the boards of both institutions.[3] The reformed development paradigm was to arrive late in Haiti, however, and to last only long enough to produce one PRSP, known by its French acronym, DSNCRP, a plan that was never implemented (GOH 2007).[4]

The process to produce the DSNCRP took place under the newly elected government of René Préval, ally of Jean-Bertrand Aristide, an agronomist by training, and a supporter of agrarian reform. Elected by both the popular base loyal to Aristide in addition to modest support from the Haitian bourgeoisie, Préval's administration would find itself at pains to balance these conflicting interests (Dupuy 2007). The DSNCRP was meant both to signify and to detail a new phase of cooperation between the Préval government and the IFIs, based upon a participatory process that promised partnership and collaboration rather than strong-arming from Washington.[5] Some have argued that the DSNCRP was not a paradigm shift at all, but rather an ultimately hollow gesture to pacify the popular elements of Préval's base.[6] Moreover, the process was far from participatory. The key section on macroeconomic policy was written entirely by the Ministry of Finance under the strict guidance of IFI economists. The relevant government ministries drafted the other chapters with more autonomy from the IFIs and with what some dubbed "participation light" from popular sectors.[7] Given this political context, and related caveats, the report can be viewed as a set of development priorities outlined by the technocrats and intellectuals favored by Préval's administration.

Given that any development plan requiring approval from the IFIs would have limited capacity to diverge radically from neoliberal goals, the content of the DSNCRP nonetheless laid out a significant reorientation of development priorities under these constraints. The document opened with a diagnostic and brief analysis of poverty in Haiti, set out three so-called priority pillars, and finally detailed actions to be taken in each of these three macro areas – growth vectors, human development, and democratic governance. In long appendices, the report elaborated detailed implementation budgets for each priority area totaling close to 4 billion dollars.

The significant difference in the report from previous plans was the identification of the country's priority "growth" vectors, albeit, and not surprisingly, still couched in dominant terms (i.e., as sources of growth) (GOH 2007: chapter 5). The report identified four: agriculture and rural development, tourism, infrastructure modernization, and science and technology innovation. The bulk of the report's attention focused upon agriculture divided into nine areas of intervention. In rural regions with high production potential (i.e., fertile plains and certain mountainous areas), the authors called for funds to increase the capital and technology inputs in production, while in areas of lower fertility, the DSNCRP prescribed subsidies to production and marketing. In all regions, the authors argued for balance between export crops and domestic food production (GOH 2007: 31). The specific recommendations sought to create the conditions to boost local production for the Haitian market, including significant state interventions responding to long-standing demands for irrigation systems, roads, and the improvement of rural markets. The authors also prioritized the development of small- and medium-sized firms for agro-processing, especially in rural areas with low agricultural potential. The section of the proposal that departed most strongly from neoliberal orthodoxy was written under the heading "promotion of new marketing strategy" (GOH 2007: 33). Here, the DSNCRP stated clearly that Haitian agriculture was hamstrung by its inability to sell products in the domestic market and called for revising tariffs "depending on whether imports are competing with local products or are convenience goods not produced in sufficient quantity domestically," as well as consultation with agricultural producers on Haiti's international trade policies (GOH 2007: 33). The inclusion of import protection for competing national products reflected long-standing demands of Haitian rice farmers who have faced near insurmountable barriers from the combined effects of state disinvestment in domestic production and the lowering of import tariffs. Moreover, the suggestion of raising tariffs on competing food crops strongly contravened neoliberal orthodoxy, even in its "poverty-reduction" mode.

The report's utter silence on garment exports, and export assembly more generally, spoke volumes. There was no mention made of trade zones or garment sector promotion as a so-called growth vector, despite the resources and legislative maneuvering on the part of the United States to promote the sector during this period (as I detail below). At the time of the report's writing, the garment sector represented 88 percent of Haiti's exports and was the country's second largest employer after the public sector (Nathan and Associates 2009: 20, 36). While the DSNCRP called for attracting investment, the authors emphasized national production, to be supported by their fourth area of focus, science and technology innovation. The report's position on the country's current export model could be read between the lines. For example, in the final paragraph of the chapter detailing the country's so-called growth pillars, the authors wrote:

> With regard to trade and development, the priority actions are: i) promotion of competitive national investment; ii) seeking accelerated export growth.

This entails, at one and the same time, favoring the emergence of *multiple and diversified initiatives* with respect to the production of agricultural and artisanal goods, manufactured products and services.(GOH 2007: 37, emphasis added)

By calling for diversified exports, the report obliquely criticized the model of garment export promotion based on US market access. If one had any doubt about the government's priorities from the text, the 12-page budget for the plan laid them out clearly: the government requested nearly 75 percent of the funds for infrastructure, 23 percent for agriculture, and 2.5 percent for tourism, commerce, and industry. The budget line to "revitalize light industry" received a request for no funds at all (GOH 2007: Pillar 1 in Appendix). The report's silences spoke volumes: Haiti, the authors spelled out in no uncertain terms to their lenders, was an agricultural country, and the key to equity in the country was to construct a viable agricultural sector.

What happened then to Haiti's DSNCRP? Why was domestic agricultural production, so clearly identified as the country's priority through the IFIs' own mandated consultative development planning process *not* the central focus of Haitian development policy in the years following the plan's publication? Why, in short, was the Caracol industrial park the flagship development project of post-earthquake Haiti?

The answers to these questions lie, I argue, in the power of the global factory both as a strategy for accumulation and as a development discourse. We will never know whether the DSNCRP would have been successful in shifting – if only slightly – more than 30 years of development policy and inaugurating a new era of state–IFI cooperation on rural development. The subsequent crises that rocked the country – beginning with a consumption crisis and devastating hurricanes in 2008, and the devastating earthquake in 2010 – buried the poverty-reduction strategy under calls for Haiti to rehabilitate its "traditional" sources of growth as the only solution to the states of emergency brought about by these waves of socioeconomic and socionatural disaster.[8] In their wake, policymakers and government leaders advanced the global factory as a development imperative, hinging existing and proposed trade zones to notions of discrete territorial progress, while re-inscribing Haiti within subordinate global relations.

## Trade Zones as Containment: The Imperative of Economic Security

In 2006, as the Haitian Ministry of Planning was developing the DSNCRP, the US Congress passed the Haitian Hemispheric Opportunity through Partnership Encouragement Act (HOPE I). The act provided duty-free access to apparel assembled in Haiti from cloth made by firms in the United States, or in countries party to a free trade agreement with the United States (e.g., the Dominican Republic), for three years. Like other similar bilateral and regional trade preferences, the legislation carried residues of the Cold War policies that initially gave rise to these trade arrangements: duty-free access to the US market was conditioned upon Haiti's adherence to free market principles, including a minimal role for the state, the elimination of barriers to US investment and trade, and national treatment for US investors (see Chapter 2).[9] Most observers of the measure agreed that its provisions were too restrictive either to attract new investment or to boost production in existing garment factories.[10] Nonetheless, the act put the legislative architecture in place that would soon propel Haiti into the position of a high-volume, low-wage assembly platform for the region, offering poverty wages that were competitive with Chinese and southeast Asian producers in the hyper-competitive post-quota global garment trade.[11]

As these trade preferences were being secured, the *denouement* of the deleterious trade and economic policies of the previous decades came to a head. Structural adjustment of the 1980s and the 1990s had left the country dangerously dependent on food imports. Indeed, US exports of rice to Haiti increased by 400 percent in the years following the mid-1990s tariff reduction (USDA 2000; Childs 2012), while Haitian domestic agricultural production suffered declines in market share and protracted disinvestment, hastening a rural exodus primarily to the capital city.[12] The influx of rural migrants, underinvestment in housing, and lack of available land contributed to sharp increases in living costs in the city.[13] Thus, when worldwide food and fuel prices skyrocketed in 2007–2008, Haitians faced a perfect storm as they found basic foodstuffs priced beyond their means.[14] The acute commodification of the means of life was eroding the possibilities for the reproduction of life itself.[15] Haitian protestors named the pangs in their stomachs "Clorox hunger" after the burning sensation caused by price-induced undernourishment. Thousands took to the streets to demand a resolution to the high cost of living from their government, eventually forcing the resignation of the Prime Minister and his cabinet, followed by the temporary implementation of subsidies (Fatton 2011a).[16]

The 2008 consumption crisis and the ensuing political turmoil, coupled with a disastrous hurricane season that devastated the northern region, provided the impetus for a new development proposal. The blueprint for such a proposal was written by Oxford economist Paul Collier in the form of a brief, polemical report titled *Haiti: From Natural Catastrophe to Economic Security* (2009).[17] Before turning to the report, it is necessary to reflect upon Collier's broader development approach, concisely summarized in his popular text *The Bottom Billion* (2007). According to Collier, while global capitalism has improved the welfare of most inhabitants in the so-called developing world, the "bottom billion," or the world's population that lives in extreme poverty, is trapped in this condition by geographical and/or political constraints. Echoing "big push" narratives of 1960s development discourse, Collier argues that the world's poor need "the West" to intervene directly and more aggressively in so-called failed or weak states in order to guarantee conditions for growth and thus to escape these development traps. For Collier, new or persistent concentrations of economic deprivation are exogenous to the global economy rather than constitutive of its uneven development. Excluded from reaping the rewards of global economic growth, the inhabitants of these margins – racialized in his text and glossed by the term "Africa +" which includes portions of the African continent, Haiti, and Central Asia -- constitute a threat to the West. Linking to the post-9/11 discourse on terror, Collier argues that the West faces a stark choice: either to act on behalf of the bottom billion or to live in fear of this excluded other. The only way to contain this threat is vigorously to create institutions to support markets, and to intervene militarily if necessary to establish these conditions. While the latter may cost "Western" lives, Collier writes, "I don't want to see [my son] exposed to the risk of being blown apart in London by ... some exile from a failing state" (2007: 126).

Collier's work represents a broad shift in development discourse and policy towards a strategic blending of economic and social development priorities with a security agenda.[18] In his report for the United Nations (UN), Collier turned this security lens on Haiti, calling for an urgent shift in planning priorities to what he deemed to be "simple and immediate objectives."[19] Haiti's social peace following the unrest of 2008 was fragile. To be secured, Collier argued, the government would have to eschew "distant grand visions," and instead focus on four priorities: job creation, basic services, food security, and environmental security. The report achieved a manifesto-like tenor of urgency and immediacy by foregrounding two administrative deadlines and one classic looming threat. Collier argued that both the UN Stabilization Mission in Haiti (MINUSTAH), a multilateral force in the country since 2004, and the newly revised trade preferences called HOPE II (see note 11), had created the conditions for investor security and market opportunity propitious for a concentrated focus on export-oriented development that would lift Haiti out its economic quagmire. Both programs were time limited, however.[20] If the Haitian government and its international donors failed to take advantage of the window of opportunity opened by MINUSTAH and HOPE II, Collier argued, then the nearly 5 billion dollars spent on the MINUSTAH force would have failed to create long-term conditions for economic growth. But time was of the essence for reasons far more elemental than these bureaucratic deadlines. Echoing the Malthusian trope of classical political economy that has long been a mainstay of development discourse about Haiti, the most imminent threat to Haiti, Collier argued, was its rapid population growth. A "youth tsunami," Collier warned, "is accelerating the process of environmental degradation and adding to the potentially explosive pool of unemployed youth." The metaphor of explosive youth as a kind of natural disaster reproduced long-standing discourses that link blackness to "uncivil" behavior. These were repeatedly drawn upon to dispossess Haitians as bearers of rights in the wake of the earthquake (Mullings et al. 2010).

Collier, of course, is hardly unique in his characterization of Haiti as a threat to be contained, and the development–security nexus advanced by his report is part of broader development trends (Shamsie 2011). Historians and anthropologists of Haiti have long observed how US and European foreign policies have consistently framed Haiti as a threat to "Western civilization," a position rooted in the early nineteenth-century metropolitan panic that a self-governing black republic would destabilize the colonial order based upon slavery (e.g., Trouillot 1995; Dubois 2012). At the height of the AIDS scare, for example, Haitians were falsely identified as a source for the epidemic, leading to the quarantine of Haitian migrants in the US-run prison at Guantánamo Bay (see Farmer 1994). As Duvalier's rule unraveled, the US coastguard implemented an interdiction policy to stem the tide of Haitian boat people arriving on the sandy shores of south Florida (DeWind & Kinley 1988). The IMF-sponsored structural adjustment policies of the early 1980s were designed in part as a response to Haitian emigration, based upon the erroneous assumption that disinvestment from domestic agriculture and promotion of cash crops and assembly production would create jobs in Haiti and thus alleviate international emigration (DeWind & Kinley 1988).[21] During this period, in Haiti as elsewhere in the circum-Caribbean, trade zone promotion was also seen as a way to contain the spread of Communism on the heels of the Sandinista revolution in Nicaragua and US containment policy towards Cuba (see Chapter 2).

In the wake of recent socionatural disasters in the Caribbean in recent years, we have seen a new iteration of containment practices and threat discourses, deeply tied to anti-black racism.[22] Following the earthquake in Haiti, for example, reports of looting and lawlessness were used to justify the US military's focus on securing key areas of the devastated region before desperately needed supplies and equipment would be delivered. Partners in Health, an international health agency founded by well-known US doctor and Haiti advocate Paul Farmer (see below), estimated that as many as 20,000 people may have been dying daily who otherwise could have been saved if disaster assistance had been the top priority (Dugan & Dade 2010;

Mullings et al. 2010). The United States also prioritized refugee deterrence despite little evidence that Haitians were taking to the sea. Within 72 hours of the earthquake, while aid provisions had yet to arrive, US planes made numerous sorties along the Haitian coast, dropping flyers and announcing in Creole to a refugee population that would never materialize not to seek relief by traveling to US shores (Padgett 2010; McKinley Jr 2010).

The Collier report both reflected and reproduced this imperative of threat containment, and drew on this security narrative to create a blueprint for an export policy little changed from that of the structural adjustment period 30 years before. In place of the DSNCRP's emphasis on bolstering domestic agriculture, one-third of Collier's report was dedicated to the promotion of assembly production in trade zones. Simply put, Collier argued that the combination of the lowest wages in the hemisphere together with guaranteed duty-free access to the world's largest market under the HOPE II legislation made Haiti an obvious pole for garment production. The global garment industry would promise practically infinite growth for the country given its small size relative to the massive garment market in the United States.[23] If export production could surpass a threshold of initial investment to create a cluster of firms, Collier asserted, the effect would be to drive down the marginal cost of production for each additional firm, and "the industry can expand until it runs out of labour."[24] This cluster argument was par- ticularly misleading since the model of export assembly is premised entirely on subsidies to the investor. The state, as well as donors in the case of Haiti, covers the major start-up costs, such as energy provision and infrastructure, not firms. Together with government pledges to sup- press wages in the sector, it is this guarantee of minimal sunk costs – and thus hyper firm mobility in the event of political instability or union organizing – that secures this type of investment. As Lon Garwood, rep- resentative for the SAE-A garment conglomerate setting up production in the new Caracol park post-earthquake, told the *New York Times*, "[t] here is a big gamble here if things were to go to hell politically ... What has helped us to mitigate the risk is that we didn't have to come in and build our own power plant. We didn't have to bring in money and buy land and bring in money and build factories" (quoted in Sontag 2012).

Repeating such easily debunked nostrums, Collier went on to assert that Haiti would simply need to set up "a few islands of excellence rather than ... trying to improve standards across the entire country." Additionally, he added, the trade zone plan was propitious because it demanded little from the local government in terms of regulatory capacity.[25] In short, Collier concluded, "the garments [sic] industry has the scope to provide several hundred thousand jobs to Haitians and to do so over a period of just a few years."

Collier's championing of the global factory as the country's top development priority is significant not only in the kinds of policies and jobs that follow, but also in the erasures and silences that are created by this dominant vision. Attention to both aspects is important. In terms of the former, the low-wage export model has been roundly critiqued by other scholars in relation to Haiti, in neighboring Dominican Republic, and much more broadly. Across the heterodox development field, there is an increasing consensus that such a singular focus on accessing the bottom rungs of the garment commodity chain is a short-term and ultimately dead-end strategy for accumulating capital to be reinvested in the country, and for creating jobs that do not treat workers as disposable.[26] Haiti's own experience in the 1970s and 1980s serves as an example. Although a supporter of the country's early export assembly industry, the World Bank nonetheless found that by the end of the industry's first decade, trade zone factories had failed both to catalyze higher-value investments and to create forward and backward linkages. Due to these limitations, the Bank concluded that over the course of the 1970s, the industry actually spent more foreign exchange on the imports of capital goods for factories and luxury consumption goods for their owners than it generated through exports (World Bank 1978; see also Dupuy 2010).

Similar to the model's early days in Haiti, contemporary promotion of assembly has been premised upon wage suppression in order to guarantee the lowest wages in the hemisphere, much touted by development planners and investment promoters like Collier and Bill Clinton. In 2009, organized labor demanded a modest increase in the daily minimum wage of 70 gourdes (1.75 dollars), a rate that had not been increased in six years and had been eroded substantially by inflation. Non-union workers joined in the demonstrations for the increase, creating a considerable mass movement. The Haitian Industry Association (ADIH) lobbied vigorously against the proposed increase to 200 gourdes (5 dollars), arguing that the increase would hurt Haiti's ability to compete in the global garment industry and cause layoffs. In fact, as industry analysts noted, the garment sector in Port-au-Prince was paying well above the minimum wage by this time, reporting average daily wages between 3.80 and 6.50 dollars (Nathan and Associates 2009). Nonetheless, the local garment owners argued that the increase would raise costs and scare off foreign investors and buyers, and their claims were bolstered by diplomatic efforts on the part of the US Embassy in Haiti and a USAID-funded study (Coughlin & Ives 2011). Seeking a resolution that would satisfy both the powerful interests of garment contractors and their US sponsors, on the one hand, and the demands of workers, on the other, President Préval created a two-tiered minimum wage: 125 gourdes (3.13 dollars) for export assembly workers – nearly equivalent to the nominal minimum wage in 1985 (Trouillot 1990) – and 200 gourdes (5 dollars) for all other businesses (Shamsie 2010). Assembly worker wages would increase to the new national standard over a three-year period.

These direct material consequences of export garment promotion must also be considered in relation to the erasures and silences effected by the global factory as a discourse of development-as-industrialization. In the case of the Collier report, short shrift was given to domestic agricultural policy. In light of the food price spikes in 2007–2008, Collier proposed that donors and the Haitian government establish an insurance scheme to be triggered by price thresholds that would supply time-limited subsidies to Haitians, allowing families to "adjust gradually" to market conditions. The section on food security – four short paragraphs in comparison to five pages on trade zones – made mention solely of food-for-work programs and infrastructure needs to facilitate the transport of inputs and products to and from rural producers.

Such a substantive policy exclusion obviously contradicted the country's own IFI-approved strategy paper, the DSNCRP. Collier went to considerable lengths to de-legitimize the government's poverty-reduction plan. He began by chastising the donor community for its participatory planning process under the PRSP regime in general. The failure in Haiti to date to turn the stability and market access afforded by the UN and the United States respectively into a humming economic recovery stemmed from "a misperception on the part of the donor community that *design* of economic strategy is solely a matter for the government" (emphasis in the original). Economic planning, Collier emphasized, was a technical skill and the Haitian government simply lacked the capacity to craft its own development plan. While not attacking the DSNCRP directly, Collier simply stated that documents "such as the PRSP [DSNCRP] … are essentially lists of projects drawn up by each ministry and then aggregated." Collier argued that as a fragile state, the government of Haiti was dependent on a host of other actors including donors, the UN Security Council, and local and transnational capital. Development planning via a PRSP had failed to reflect the priorities of those other actors – in particular, local and transnational capital. Again, invoking the tenor of urgency and fear, Collier wrote, "any action has to contribute immediately to building economic security rather than merely build the foundations for some distant goal." When it came to trade zone production, the required development planning had to be responsive to the demands of foreign investors and lead firms. "[I]t is essential to work back from a benchmark of global competitiveness and determine what standard must be met," he argued, "rather than working forward from any existing plans."

In short, the twinned socioeconomic and socionatural crises of 2008 created the opportunity for an emboldened policy plan framed within a security paradigm that repositioned the global factory as the only rational development approach to stave off instability and economic degradation. The priorities of the Collier report were soon validated by a new government plan that similarly positioned trade zones at the center of Haiti's development priorities (GOH 2009). In the wake of the 2010 earthquake, Collier's short proposal – with its language of urgency, security, and threat – served as a key pillar for the government's reconstruction proposal, dubbed by some as a "Marshall Plan" for the country (Dupuy 2010; Heine & Thompson 2011). Thus, similar to Naomi Klein's (2007) notion of disaster capitalism, socionatural disasters and civilian protest in 2008 gave rise to a crisis discourse that reset development priorities in line with those of transnational capital and the US government. Collier's report, however, and the general thrust of reconstruction priorities that followed, did not signal the triumph of market-led regulation over an entrenched welfare or Third World developmental state. Haiti's economy had already experienced radical structural reform in the 1980s and mid-1990s and could well be one of the most "liberalized" and "privatized" nations in the hemisphere, if not the world.[27] Rather, humanitarian and political crises in the Haitian case displaced the relatively modest proposal to "socialize" or "embed" neoliberalism under the poverty-reduction framework, reintroducing export zones as so much old wine in new bottles.

## From Poverty Reduction to Poverty Redistribution? Decentralization after the Earthquake

If the Collier report resorted to the abstract language of containment in "islands of excellence," the location of trade zones post-earthquake added an extra dimension to the security imperative that I have traced thus far. Following the quake's devastation of the capital city, trade zone promotion was hinged to long-standing claims surrounding uneven economic and political development in the country, revealed in stark relief by the disaster. Standing in the Rose Garden next to President Obama two months after the earthquake, President René Préval told the esteemed crowd that to rebuild Haiti, donors and the government had to agree on an effective decentralization policy. Such a policy required "offering health care, education, [and] jobs to all Haitians, men and women, regardless of where they live," so that fewer people would be compelled to migrate to the capital (Préval 2010). The call for decentralization as the guiding vision for Haitian reconstruction was taken up by powerful elites in the country and subsequently reflected in the 55-page Action Plan for National Recovery and Development (PARDN, in French) prepared by the government, with IFI sponsorship, for the March 2010 donors conference (GOH 2010).[28] "Reconstruction must take place elsewhere," the authors of the PARDN wrote, "at least in part" (GOH 2010: 16). The government proposed four priority growth regions where officials hoped to stimulate economic activity and extend services. While the PARDN called for support for diverse sectors, including agriculture, trade zones were given a prominent position as drivers of decentralization. The government, for example, pledged to "intervene directly whenever necessary to implement the necessary infrastructure [for trade zones and industrial parks] and ensure more balanced geographical distribution for job creation" (GOH 2010: 17).

The notion of decentralization has a particular history in Haiti that long precedes the earthquake and the recent policy debate.[29] Haitian government plans post-earthquake referenced the mandate for decentralization enshrined in the country's 1987 Constitution, which set out the legal framework for democracy after the 29-year patrilineal dictatorship under the Duvaliers. The Constitution called for a significant shift in the representative apparatus of the state and in the flows of power and wealth that had historically concentrated in the capital. While such policies were never implemented, their promulgation in the Constitution reflected the spirit of Haiti's post-dictatorship transition. What peasants and other provincially based movements were demanding was the reversal of a century-long political and economic dynamic that had concentrated wealth and centralized commerce in the capital city, while at the same time aggressively dispossessing much of the rural population. This political process of weakening provincial power was hastened by the US occupation (1915–1934), in an effort to exert control and pacify the largely rural-based resistance movement during that period. Under the Duvaliers, two decades later, the primacy of Port-au-Prince continued unabated. In the 1970s, numerous observers of the country's export strategy noted the negative effects of factory concentration on the capital city, which grew to nearly one million inhabitants over the course of the decade. Job creation in export zones could not keep up, and the unplanned and non-serviced neighborhoods of Cité Soleil, Martissant, and Bel Air swelled with the ranks of the under- and unemployed.

Post-earthquake, donors and the Haitian government appeared to agree that the supposed ailment – concentrated factories – could be transformed into the cure for unbalanced territorial development. The IDB, a major provider of financing aimed at stimulating the private sector, framed its interventions in relation to the problem of concentrated growth, noting that "past economic policies have concentrated

80% of all industrial, commercial and banking facilities, 66% of GDP and more than 90% of the banking sector loan portfolio [in the capital]" (IDB n.d: 2). The Bank's strategy was peppered with language on the need to alleviate "regional imbalances," to seek "spatially equitable distribution of economic activities," and to foster regional development poles (IDB n.d.: ii, 2, 3, 6). Its flagship private sector development project was what the Bank called the Northern Economic Growth Pole, anchored by the Caracol industrial park and the SAE-A garment conglomerate. For the IDB, promoting private sector investment in regions outside the capital had become the key policy solution for the problem that decentralization policy was meant to address. Since the earthquake, the Bank had tripled its private sector development staff in its country office, and promoted concessional loans to at least three garment factories, in addition to the infrastructure support for Caracol. The Bank's private sector lending extended well beyond garment factories, however, and included loans and grants to promote agro-exports and mining in the country, financing multinational interests directly or indirectly, as well as loans directed towards small enterprises to service these new large-scale ventures.[30]

As such policies lay bare, the vision for decentralization post-earthquake fails to specify correctly the imbalances that led to concentrated development in Port-au-Prince. It was not the location of the trade zones that contributed to the growing misery of the capital, but the low wages paid to workers, the drain on state revenues, and the shallowness of the investment, which failed to produce linkages with other local businesses. In other words, the problem was not that over 100 factories concentrated their operations in Port-au-Prince. The problem was that the dominant social relations and associated state actions created the conditions that allowed these factories to pay poverty wages, contribute little to the fiscal budget, and to expatriate all profits. In the name of spatial decentralization after the devastation in Port-au-Prince, business leaders in the country appeared to be making a similar bargain with transnational capital, backed by the Haitian state. Their proposals seemingly ignored the history of export assembly in the country, and thus risked a repetition of similar mistakes, albeit with a new spatial form. Given these conditions, the investment and private sector promotion projects for the northern region likely promised the spatial redistribution of poverty, as opposed to substantial reduction. As Gillian Hart (2002) notes in her work on rural industrialization in South Africa, however, the outcomes – still unknown at the time of this writing – were sure to be shaped by the particular social relations and histories of the regions where these plans were being implemented.

## Making Space for Development Alternatives

My analysis of development plans before and after the devastating 2010 earthquake in Haiti demonstrates the power of the global factory, not only as a material site of exploitation and accumulation, but also as a producer of meanings through erasures and silences. In Haiti, crisis-induced promotion of the global factory sought to transform complex, nonlinear, and fundamentally sociospatial histories of connection into narratives of industrial transition traversed by individual nation-states. What is apparently surprising about the global factory as a transition discourse is its power in the contemporary period, more than three decades after structural adjustment, and in light of the limitations of export assembly expounded by major development institutions, as I argued in the Dominican case in Chapter 3. For Haiti "traversed the stage" of export manufacturing in the 1970s and 1980s with dismal results that furthered the conditions responsible for the country's current vulnerability to price shocks, earthquakes, and hurricanes. Nonetheless, the power of Eurocentric thinking and the interests of multinational capital reproduce this history as a discrete failure of the country, rather than the result of pernicious integration into the uneven geographies of capitalism. In report after report, donors thus continued to frame "the problem with Haiti" in remarkably consistent terms, as one of a lack of exposure to, or exclusion from, global markets.[31]

Following the earthquake, one of the foremost critics of the export assembly model of the 1980s, Paul Farmer, changed his position on the question. In the 1980s, Farmer, an anthropologist, medical doctor, and Haiti advocate, forcefully critiqued these operations as contributors to, rather than alleviators of, the grinding urban poverty of Port-au-Prince (1988). Over decades, he pioneered an alternative community health project that grew into a large, international NGO called Partners in Health/Zanmi Santé. Farmer understood the urgency of job creation to sustain any advances he and other medical staff could make in rural health. As a result, he became an advocate for, and subsequently implemented, a system that employed paid community health workers, who numbered around 4,000 at the time of the earthquake.

After establishing a close relationship with former US President Bill Clinton, initially serving under Clinton during the latter's tenure as the UN Special Envoy for Haiti, Farmer appeared to back down from his earlier criticisms of the export assembly model. "Jobs outside the agricultural sector are urgently needed," he wrote in the US progressive magazine *The Nation* in 2008, following the hurricanes and the extension of the HOPE Act. "This should make progressives slow to disparage new jobs in the tourist and apparel industries," he continued. Farmer was subsequently criticized from the US left for his tacit support of the dominant reconstruction model. In his recently published book, he responded to these criticisms. Farmer argued that while massive investments in domestic agriculture would do much more to alleviate poverty than thousands of low-wage factory jobs, these two strategies were mutually necessary. "[D]id development need to be a competition for scarce resources when all other parts of the Haitian economy needed investments too?" asked Farmer (2011: 34). Clearly improving conditions for smallholder farmers also involved creating job opportunities for the many who could not be supported by agriculture.

Farmer's point is an important one: no matter what kind of support for domestic agriculture can be marshaled, other alternatives are necessary. The reasons are multiple, and cannot be reduced to a narrow economism. As I discuss in Chapter 4 and other ethnographers have long noted, "development" is not solely a dominant discourse wielded by international NGOs and donors. For many people, especially younger generations and women, development means leaving agriculture and the attendant constraints of agrarian social relations in order to partake in an urban "modernity." From the perspective of new migrants and workers negotiating livelihood strategies under conditions not of their own choosing, factory jobs may offer a (likely short-term) solution, but one that often comes at a high cost to workers' long-term health and well-being.

Apart from these points, which I have covered extensively throughout this book, the answer to Farmer's question of whether the dominant export model must be seen to be in competition with support for other sectors is, unfortunately, yes. The global factory as a development discourse normalizes a sequentialist, Eurocentric understanding of development, at the cost of rendering Haiti's long and complex history with the model marginal or insignificant. While the energies and resources of the largest donor organizations are channeled into the projects that signal such "progress," this orientation towards export assembly is also reinforced by the bilateral trade preference regime. If history is any indication, such regimes bolster narrow interests among the exporting elite, who consequently work against efforts to transform the low-wage, export-oriented model into a more sustainable one, capable of redistributing wealth through living wages and by creating space for local production (Cox 2008). My point is not that development is a zero-sum game where resources that go into a certain sector do so at the expense of others. Rather, the dominant development model – represented by the global factory – makes the prioritization of progressive land reform and smallholder agriculture for domestic consumption over factory wage jobs and export cash crops unthinkable. Barring widespread social demands for reform, the dominant model reflects and reproduces existing social relations, both national ones and those that position Haitian farmers and workers in subordinate positions to transnational capital and the US government.

I was reminded of this quandary in my interview with the head of the Haiti office of the IDB in 2012.[32] The official described the Bank's projects to create the Northern Economic Growth Pole with great enthusiasm, and predicted that 150,000 jobs would be created, just over half in the export manufacturing sector anchored by the Caracol industrial park. The pole was much more than the trade zone, he assured me; it was an integrated regional plan. The plan included a tourist circuit connecting the Labadie bay, already a concession to Royal Caribbean cruises, with the historic citadel, from whence the ill-fated Henri-Christophe once ruled the north of the country. The Bank was working with regional planners to create housing in towns near the industrial park in an attempt to preempt over-crowding and slum formation around the trade zone. Finally, the IDB was financing value chain projects in agriculture to substitute some of the imports coming from the Dominican Republic, like bananas. "There's a big effort to make the pole a real model so that we can first of all create the 150,000 jobs," he said. "If you do that," he continued, "you are going to create a new economy." I asked how the project's implementation was going. The official explained that they had identified 1.1 billion dollars for investment. The government, the IDB, and other donors, he told me, had already financed about half of that amount. The road, for example, connecting the cruise ship concession to the citadel had financing. It was more difficult to find financing for other parts of the project, however. "The secondary roads for agriculture, for example, we still haven't found the institution to finance those," he explained. Not surprisingly, the badly needed infrastructure to strengthen local economic activity in the hands of Haitian farmers was of second-order importance.

One cannot predict how local conditions, especially the local social relations uniquely shaped by the particular history of Haiti's northern region, will transform the planned development of the area. By 2013, workers from the CODEVI trade zone in Ouanaminthe had traveled to the new trade zone to organize Caracol workers just as the factories were starting operations; Caracol workers had subsequently registered an independent union within just a few months of the park's opening. The degree of success that these organizations will have in wresting a portion of profit from US retailers and brands in order to transcend their status as disposable workers cannot be foretold. The question of development alternatives remains, however. In the chapter that follows, I draw upon a variety of responses to multiple rounds of economic crisis and structural adjustment in the Caribbean to explore practices that contest the narrow and disabling imaginary of development crystallized in and through the global factory.

# Notes

1   The Caracol park was extensively covered in the press, and not always favorably (e.g., Charles 2012; Sontag 2012). The inauguration of the park's operations in October 2012 was an even more high profile affair than the groundbreaking, featuring Hilary and Bill Clinton, Donna Karan, Ben Stiller, and Sean Penn, along with Daniel Cho and other representatives from SAE-A. Key investors in Haiti and/or the Caribbean were reportedly also on hand, including Richard Branson of the Virgin Group, Dennis O'Brian of Digicel, currently Haiti's largest private sector employer, and major Dominican moguls Alfonso Fanjul of Florida Crystals and Frank Reineri of Punta Cana (IDB News Releases, October 24, 2012).

2   For a broader discussion of post-Washington Consensus policy, see Rodrik (2006) and Serra and Stiglitz (2008). For a discussion of the internal tensions between the World Bank and the US government (specifically the Treasury) over poverty reduction in relation to redistribution and economic growth, see Wade (2002). For critical discussions of the resurgent poverty discourse in development under Wolfensohn and Stiglitz at the World Bank and the Millennium Development Goals of the United Nations, see Cammack (2004) and Roy (2010). See also my discussion in Chapter 3.

3   The PRSP process was introduced under Wolfensohn at the World Bank in 1999. There is a large literature on this topic. For critical assessments of the PRSP process, see Lazarus (2008), Craig and Porter (2003), Abrahamsen (2004), and Cammack (2004).

4   The PRSP process was rolled out worldwide during a period when Haiti was increasingly shunned by the donor community as an unruly aid recipient, particularly once George W. Bush took office in the United States, culminating in a three-year aid embargo that contributed to destabilizing the Aristide government (see Chapter 5).

5   Indeed, this process began following the 2004 coup against Aristide. The transitional government and the donors produced an Interim Cooperation Framework (UN 2004), signaling a rapprochement between the two sides now that Fanmi Lavalas, Aristide's political party, was out of power. The framework claimed to be a "break with the past," recognizing the failures of past aid and pledging "a joint and realistic identification of needs" by both the government and the donors (UN 2004: xi), and the construction of even broader participation in the forthcoming PRSP process. Nonetheless, popular groups viewed the ICF as a neoliberal pact with Washington and, once Préval was elected in 2006, encouraged him to break with the framework (Dupuy 2007).

6   Personal communication with Alex Dupuy, June 14, 2012.

7   Interview with Hérard Jadotte, former coordinator of the technical secretariat for the preparation of the DSNCRP, Ministry of Planning, Port-au-Prince, July 10, 2012. See also the thesis by Pradhally Nicolas, titled "Participation paysanne dans le processus d'élaboration et de mise en oeuvre du DSNCRP: pour réussir le saut qualitatif?," July 2012, on file with author.

8    In fact, the designation of "tradition" here represents a problem for donors and development planners who seek to advance an export-led growth strategy. Indeed, after 40 years, light-goods assembly can hardly be designated as a new economic sector. Yet, the need to frame the industry as a sign of progress can lead to awkward designations of textile exports as at once a traditional sector, and, at the same time, part of a strategy to pursue non-traditional exports (e.g., IDA & IFC 2011: paragraph 17).

9    Under eligibility requirements, for example, Haiti must be certified by the US president as "a market-based economy that protects private property rights, incorporates an open rules-based trading system, and minimizes government interference in the economy through measures such as price controls, subsidies, and government ownership of economic assets" (HR 6142).

10   US buyers and Haitian and Dominican producers criticized the initial HOPE legislation for being too restrictive: the three-year time horizon did not provide enough of an incentive for producers to export under the scheme; the eligibility requirements for value added were too high; and the requirement for direct shipment from Haiti barred producers that were using Dominican port facilities from eligibility. The legislation also provided for so-called trade preference levels (TPL), or a quota for duty-free treatment for a given amount of woven apparel produced from third-party fabric (i.e., cheaper fabric produced in Asia). Such quotas faced stiff opposition from US textile producers since they eroded the protection that these preference schemes ultimately granted to US textile producers. In addition, advocates of the HOPE Act argued that the TPL quota was too low and, because it only applied to woven garments, would only have limited applicability in Haiti since producers there mostly made knit garments (Hornbeck 2010: 17).

11   Very few producers exported under the provisions of HOPE I. As a result, the US Congress amended the act in 2008, creating HOPE II. Among the major changes, the time horizon for preferences was extended an additional 10 years, rules of origin were loosened, the TPL quota (see note 10) was raised, qualifying goods could ship from the Dominican Republic, and an allowance was made for producers to qualify for additional exports from non-qualifying cloth (i.e., largely from Asia) at a given ratio. Finally, US legislators included enforceable provisions on labor rights in the act in response to criticisms from international and US unions, and established a monitoring program run by the International Labor Organization (ILO) called the Better Work Programme, modeled after an arrangement put in place under a similar apparel preference scheme with Cambodia in the late 1990s. This trade preference architecture was further revised following the earthquake under the Haiti Economic Lift Program Act of 2010 (PL 111-171), which extended the preference scheme until 2020. See Hornbeck (2010: 18–23). The ILO country program, called Better Work Haiti, was initially funded by the US Department of Labor, which also oversaw the program.

12   This process has been extensively documented. See for example, Trouillot (1990) and DeWind and Kinley (1988) for an analysis of early structural adjustment and the dispossession of rural dwellers. See Weisbrot (1997) for a discussion of the mid-1990s reforms.

13   The combined effects of inflation and wage stagnation created declining living standards for trade zone workers. For a retrospective testimony on this decline, see the documentary *Poto Mitan* (2010), especially the narrative of Thérèse (http://potomitan.net/therese.html).

14   Worldwide, the cost of rice increased by 74 percent between March 2007 and March 2008 (McMichael 2012). This increase contributed to what had been a steep erosion of Haitian's purchasing power, which declined by 28 percent between 2003 and 2008 according to official estimates (ECLAC 2010). From January to April 2008, a 55-pound bag of rice went from 13.90 to 28.20 dollars; 50 kilograms of sugar rose from 31.50 to 44.70 dollars and 50 kilograms of flour rose from 31.50 to 53.00 dollars (Fatton 2011a).

15   See Polanyi (1957 [1944]) for a theoretical discussion of the biosocial effects of commodifying the means of life, bringing about a "double movement" to socialize the market.

16   Préval mandated a 15 percent price reduction in the cost of rice and major international donors financed short-term relief (Fatton 2011a).

17   At the time, Collier was a special adviser on Haiti to the UN Secretary-General. Collier served as the head of research at the World Bank under Joseph Stiglitz during the latter's tenure as lead economist. Like Stiglitz, he is closely associated with the post-Washington Consensus reformist turn in development policy, which broadly argues for policies to support institutions "to make markets work." These can be thought of as second-generation reforms, following on the heels of structural adjustment policies of the 1980s and 1990s. See also note 2.

18   See, for example, the World Bank's 2011 World Development Report on Security. For a detailed analysis of the development–security nexus and its trajectory in Haiti, see Shamsie (2011).

19   All direct quotes are taken from Collier (2009); the report is not paginated.

20   The MINUSTAH mandate and the time horizon for trade preferences were both subsequently extended following the earthquake.

21   Sassen (1988) debunked these assumptions early on, albeit in an economistic way, by demonstrating that export assembly jobs were accelerators of emigration as workers used these jobs to leverage resources to emigrate. The deep and complex connections between export assembly and international migration were traced in this early period by Fernández-Kelly (1983). More recently, they have been well documented by Kelly (2009) in the Philippines and Mills (2005) in Thailand, among many others.

22   There are striking parallels between the focus of media attention on criminality and the violent efforts to contain the largely African-American population of New Orleans following Hurricane Katrina and the press coverage and military policies implemented in Haiti after the earthquake. For an exploration of these issues, see Mullings et al. (2010).

23   It is worth noting that growth of consumer markets in the global North has not increased at the same rate as productive capacity, intensifying the industry's hypercompetitive conditions and the turn to strategies like "fast fashion" that expand the market by intensifying product turnover time.

24 The statement echoes the foundational theories of development economics, in particular W. A. Lewis' notion of economic development with "unlimited supplies of labor" and the dual economy thesis. See my discussion of Lewis' dual economy idea in Chapter 3 and Escobar (1995).

25 Writing before a spate of factory fires and the tragic collapse of Rana Plaza in Dhaka, Bangladesh that killed 1,129 garment workers in 2013, Collier cited Bangladesh approvingly as a model for Haiti, where, despite a "corrupt and inefficient government," Bangladesh had supported a sector that had grown to more than 2.5 million garment jobs by 2009.

26 See, for example, Kaplinsky (1993; 2000) and my discussion in Chapter 3 of the resignification of this model in the Dominican case.

27 By the late 1990s, the IMF determined Haiti to be the least trade restrictive country in the Caribbean. The country's tariff on rice imports, for example, had been reduced to 3 percent (from 35 percent), compared to the Common External Tariff for rice for the Caribbean Community of 25 percent (Fatton 2011a). The government operated only 8 percent of primary and middle schools (Demombynes et al. 2010). The last of Haiti's remaining state-owned industries was privatized in the 1990s (Weisbrot 1997).

28 For more discussion and background on decentralization as a political demand, and its mobilization by powerful groups and donors post-earthquake, see Elie (2011).

29 There is a large literature on this topic. I draw on Elie (2006), Anglade (1982), and Trouillot (1990).

30 The IDB has a history of financing agriculture and has approved projects to restore watersheds and to rebuild irrigation infrastructure in Haiti's key agricultural areas. Much energy and funds in agriculture, however, is dedicated towards so-called private sector development projects aimed at integrating Haitian farmers into global export markets – in mango and coffee, for example – and into input markets controlled by large multinationals like Monsanto. See Shamsie (2012) for a discussion of agricultural policy in Haiti; see McMichael (2013) for an analysis of the recent proliferation of these "value chain development" projects in agriculture.

31 The Collier report largely frames the problem as one of exclusion from global markets. The IDB, on the other hand, casts the problem as an absence of markets (e.g., IDB n.d.: 1).

32 Port-au-Prince, July 13, 2012.

# 7

# Unsettling Dominant Crisis Narratives of the Caribbean

This chapter considers the political and ethical dimensions of long-standing and emerging patterns of uneven development at the center of my study of the global factory. I approach this task through an examination of the notion of crisis, specifically the way in which crisis claims enable the Eurocentric development narratives that are at the heart of my critique. If the development discourse of the global factory reduces geographies of uneven development to a temporal sequence of stages, dominant narratives of crisis provide "events" that organize this sequencing, from socionatural disasters like the Haitian earthquake to debt crises that iteratively spark labor devaluation. These dominant crisis events, I argue, serve as a temporal mechanism that translates sociospatial change into sequential teleological narratives. Therefore, if we are to recuperate a non-teleological study of the global division of labor, dominant narratives of crisis must be challenged.

A major contributor to dominant crisis narratives in recent times, and a good starting point for an examination of what such narratives render thinkable and unthinkable, is the 2007–2008 financial meltdown sparked by the US housing market bubble. In her short polemic published in the wake of the financial crash, anthropologist Janet Roitman argues that scholars and experts of all stripes have framed the problem of "the crisis" in terms of what she calls a "sociology of error" based on the foundational question of "what went wrong" (2014: 9–10). Neoclassical accounts locate error in the price mechanism, while neo-Keynesians identify institutional market failures, and Marxists point to the deviation of financial values from material ones. In short,

*Global Displacements: The Making of Uneven Development in the Caribbean*, First Edition.
Marion Werner.
© 2016 John Wiley & Sons, Ltd. Published 2016 by John Wiley & Sons, Ltd.

the crisis narrative constructs a norm, framing how things are as a kind of failure in relation to how things ought to be. In so doing, crisis "regulates narrative constructions [and] allows certain questions to be asked while others are foreclosed" (Roitman 2014: 94).

Roitman posits how this normative framing of crisis is achieved. She argues that crisis narratives signify what counts as an event and thus establish "a temporality upon which one can act" (Roitman 2014: 7). This prognosis of time, this framing of a temporality upon which to act, is achieved by constituting an authorial "blind spot" from which to judge what counts as history – in other words, what counts as the series of crisis events that we call history. Dominant narratives of crisis, Roitman argues, thus render impossible a range of questions – and, thus answers and actions – that defy the structure of thought instantiated by this dominant framing. For Roitman, the important questions that are outside the domain of possible interrogation are those that can lay bare the conditions of possibility for producing global credit and debt markets themselves, because crisis analytics take these markets for granted in framing them in terms of failure. Academic and policy discourses about the 2007–2008 financial crisis, for example – including those that espouse a critical position vis-à-vis capitalism – are restricted to asking questions such as "who bears the brunt of the crisis." The problem is that such questions already accept the claim of crisis and its fundamental supposition of a crisis event as a kind of deviation from how things – the market, value – ought to be.

Indeed, crisis narratives are not merely descriptions of history; they are productive of certain understandings of history. And while I do not share Roitman's commitment to an alternative account that prioritizes technologies of market-making while marginalizing more material approaches, her provocation is nonetheless an important one. The departure point for my argument is that abandoning the notion of crisis is not possible (as Roitman at times implies); crisis narratives are not a matter of election. If, however, crisis narratives achieve their dominant, ordering effects by framing certain particular temporalities while foreclosing others, then in order to unsettle crisis, we must recuperate alternative temporalities and spaces so as to de-center crisis as seen through the lens of events sanctioned by Eurocentric history. In short, I propose a feminist and postcolonial politics motivated by situated understandings of crisis that disrupt Eurocentric capitalist temporalities of history and the event (De Angelis 2007). In what follows, then, I offer the reader three accounts that aim to unsettle particular crisis events in the Caribbean: the debt crisis and structural adjustment of the 1980s and early 1990s, the Dominican financial crisis of 2003, and the contemporary debt crises facing several Caribbean states and Puerto Rico since 2008. I argue that we can interrupt these dominant crisis narratives through alternatives that posit a locus of enunciation from the global South (Mignolo 1995) and foreground a feminist commitment to the Others of dominant economic discourse such as informality, social reproduction, and the time of the everyday. In the conclusion, I consider what these interruptions offer us for rethinking ethical engagements with the global factory.

## The Everyday and the Event of Crisis in "A Small Place"

Antiguan poet and author Jamaica Kincaid published a short collection of essays, *A Small Place* (1988), in the wake of the crippling debt crisis that gripped Latin America and the Caribbean in the 1980s. The first essay, a poetic denouncement of both structural adjustment and the tourism industry as forms of neocolonialism, is narrated by Kincaid herself in the widely viewed documentary on the impact of the debt crisis in Jamaica, called *Life and Debt* (dir. Stephanie Black, 2001). Kincaid's essays communicate a more fraught politics than those captured by the film, however. Her evocation of "a small place" at the center of her text vividly conveys the *aporia* of postcolonial history and sovereignty. We can understand the term *aporia* to describe an unavoidable dilemma that prevents the resolution of problems that emerge in the wake of colonialism, while simultaneously creating other political opportunities and possibilities. The small-ness of place in Kincaid's text does not simply describe the physical size of a Caribbean island, although this too figures in her narrative, but rather the structural, historical, and political limitations of "the people in a small place" to narrate their history in dominant terms that could change their relationship to that history. Kincaid's narrative combines sarcasm, irony, frustration, and no small dose of disillusionment to capture this dilemma:

> For the people in a small place, every event is a domestic event; the people in a small place cannot see themselves in a larger picture, they cannot see that they might be part of a *chain* of something ... The people in a small place cannot give an exact account, of themselves. The people in a small place cannot give an exact account, a complete account of events (small though they may be)(1988: 52–53).

To concatenate time into a narrative, a chain, a series, what is needed are "events." To determine these events and thus a history, "the people in a small place," Kincaid tells us, must be able to judge, to question, to consider in order to define and to distinguish events from the everyday in terms that give them license to narrate history from the privileged outside position of authority sanctioned by crisis. Indeed, the etymology of "crisis" in ancient Greek means to judge, to separate, or to choose. To draw again on Roitman, what Kincaid appears to be lamenting is the denial of the authorial "blind spot" that dominant crisis narratives enable precisely to judge, to choose, to determine what counts as history (Roitman 2014: 18). Yet, "people in a small place" – those forging livelihoods in postcolonial contexts constrained by the legacies of colonialism and contemporary capitalism – are constituted within an epistemological position that precisely denies this subject position from which the "events" of sanctioned – that is, Western – history are determined.

Kincaid describes this dilemma of sanctioned history and its subjects as the experience of a continual shifting between the event and the everyday. The event appears hurtling towards "people in a small place" and is experienced initially as an immense burden "until eventually," Kincaid writes, "they absorb the event and it becomes a part of them, a part of who and what they really are, and they are complete in that way until another event comes along and the process begins again" (1988: 53). Kincaid's notion of the absorption of the crisis event, its veritable neutralization into an everyday occurrence, recalls the many feminist accounts of the 1980s that documented how households and communities in Latin America and the Caribbean ultimately served as the shock absorbers of austerity (e.g., Safa 1995a). Kincaid writes that the absorption of burdensome events into everyday practices substitutes for contextualizing experience within a set of larger events – something Kincaid longs for but posits as impossible. She yearns for the non-elites of her native country to have "a different relationship with the world, a more demanding relationship," a relationship that would liberate the people in a small place from being subjected to "every bad idea that flits across the mind of the world" (Kincaid 1988: 56–57). Her lament is ambiguous and does not offer an easy reading for she could be understood either to be mocking or echoing the colonial denial of subject-hood to colonized Others by denying their "rational" relationship to time.

Kincaid's proposition of the slippage between the event and the everyday, I argue, can be read instead as signaling the dilemma of the Caribbean. She frames the im/possibility of subalterns appropriating the dominant, Western temporalization of history that has continually marginalized and undermined postcolonial self-determination. We can draw a useful parallel to Haitian anthropologist Michel-Rolph Trouillot's (1995) powerful depiction of the Haitian revolution as an unthinkable event. In realizing a self-governing, independent nation ruled by former slaves, the Haitian revolution challenged the very ontological order of the West and thus could not be thought within the West's own terms. The evidence of this fact is strewn throughout the historical record of the late eighteenth and nineteenth centuries. Trouillot finds a myriad of rationalizations advanced to interpret the upheavals in colonial Saint-Domingue through a Eurocentric lens, such as the many theories developed to explain who – the British, mulattos, royalists, republicans – stoked the uprising, for slaves could not possibly have conceived and led the revolution themselves. Buck-Morss summarizes the dilemma of black sovereignty during that period as "the paradox between the discourse of freedom and the practice of slavery": the metaphor of freedom became the *sine qua non* of Enlightenment thought and circulated via an economic system based upon the enslavement of non-Europeans (2000: 822; also Trouillot 1995). Examining the historiography of the Haitian revolution in the twentieth century, Trouillot writes, "the revolution that was unthinkable became a non-event," barely worth a paragraph in surveys and textbooks of modern history (1995: 98–99). Trouillot's and Kincaid's distinct engagements with "the event" as an *aporia* of postcolonial Caribbean history lend additional meaning to the oft-quoted observation by C. L. R. James. "The Caribbean," James wrote, "is in but not of the West" (cited in Hall 1996: 246). The region is ineluctably tied to the simultaneously revolutionary and profoundly exclusionary project of Western modernity, and systematically reduced and marginalized in any account of that project.

Let me offer an additional reading of the dilemma of crisis and postcolonial history based upon Kincaid's evocative text. Antigua's central public library was destroyed in an earthquake in 1974. More than a decade later, Kincaid writes, the ruins serve as the headquarters for a carnival troupe while a small, temporary collection is housed in a room above a store front on the main commercial street of the capital, and the majority of the public book collection rots in storage (1988: 41–52). Lamenting the loss of the library, a place that she identifies as pivotal in her own intellectual and political development, Kincaid also sketches out the limits of a rebuilt library in restoring the possibility of narrative coherence to her native land, even as she travels around the capital to advocate for funds to reconstruct the building. The library would house its complete collection of books, yes, but these are mostly about England and its empire; the restored library collection would reenact the violence faced by post-emancipation societies forced to adopt the language of their masters; and any restoration would be paid for out of a sense of *noblesse oblige* by local white elites. Kincaid's narrative eloquently conveys the *aporia* of the Antiguan central library. The library is, at once, a site constrained by the colonial relation and the subject of a text that demonstrates the unpredictable outcomes of such spaces.

Kincaid continues to canvass for support of the project that is so charged with these contradictions. Returning to the irresolvable dilemma of appropriating the dominant temporality of the everyday and the event sanctioned by crisis, Kincaid writes that such an operation "would demand the invention of a silence, inside of which these things could be done" (1988: 53). And while Kincaid posits this silence – from within which subalterns could seize the authorial blind spot in order to narrate a dominant history – as impossible, the library campaign suggests that other possibilities exist for creating spaces within which narratives that challenge Western hegemony, like Kincaid's own, can be constructed.

## Forging Livelihoods Out of Crisis in the Dominican Republic

The experiences of economic crisis and structural adjustment in the Dominican Republic live on in the narratives and livelihood strategies of the households that absorb these crises into the everyday, to deploy Kincaid's metaphor. In my conversations with former trade zone workers, or *zoneros,* the destabilizing effects of such crises were seamlessly interwoven into the narratives of their work histories and their practices to secure positions of social worth. Juan Luis, a veteran *zonero*, offers one such narrative.

I first met Juan Luis at a union meeting before the closure of IA Manufacturing where workers were discussing how severance would be paid out with their union representative. Over the next months, we met several times to talk about his work history and how things were going since the factory had been shuttered. Juan Luis was born and raised in Santiago. When I asked him about his experiences working in the trade zone, he began by telling me about his training as a shoemaker. "I remember it as if it was today," he told me in a deeply nostalgic tone. He was eight years old and he and a friend decided to go looking for work, wandering far from home and returning to find his mother very upset that the two boys had run off without permission. Despite her fear and anger at her son's disobedience, Juan Luis' mother gave him permission to apprentice with the shoemaker he had met on his outing that day. By the age of 12, Juan Luis could make a complete pair of shoes. He continued working in his spare time in the small shop of the man with whom he had apprenticed. Before starting his second year of high school, he left school and the small shop to work full time in a larger workshop. "You could make a living from shoes," he told me, "but when the flea market was opened up, it became impossible. The government doesn't invest in the sector because the flea market is everywhere."

The flea market was the outcome of new policies implemented by then-Dominican president Joaquín Balaguer. The conservative Balaguer government had initially opposed a second round of market reforms when first elected in 1986, following the collapse of the previous government whose accession to structural adjustment policies prescribed by the IMF had sparked some of the most violent uprisings in the region (see Chapter 2). Holding the IMF at bay for a few years, Balaguer stimulated the economy through infrastructure spending fueled by monetary expansion, actions that also triggered a depreciation of the Dominican peso and inflation. In 1990, faced with the external shock of increased oil prices brought about by the onset of the first Gulf War, the country could no longer finance its import bill and had to turn to the IMF for fiscal support (Betances 1995). Tariff liberalization was a condition of the multilateral fund's bailout package (Betances 1995; Isa Contreras et al. 2003). With the subsequent opening of the Dominican domestic market to shoe and clothing imports, employment in both domestic sectors collapsed quickly.

I was curious whether Juan Luis' work experience directly mirrored these macroeconomic changes and asked him when exactly he had stopped working as a shoemaker. "Really, I've never stopped," he told me, "because you never forget what you learn." In his response, Juan Luis implicitly rejected my assumption about the effects of the 1990 economic crisis and subsequent trade liberalization on his work; instead, he asserted that his skill as a craftsperson could not be completely taken from him. In the early 1990s, the shop where he worked did indeed close, and Juan Luis went to work in the trade zones, first for the US-owned shoe company Timberland, doing a single operation in the assembly system. After losing his job in a round of layoffs there, he switched over to the garment sector and got a job at the large, Dominican-owned contract firm IA Manufacturing, where his wife was already employed. She immediately taught him an operation: cutting extra cloth after hemming (*cortar ruedo*). But Juan Luis never spent more than two consecutive years at the garment plant, leaving and coming back three times over the next decade. He would return to shoemaking in small street-side workshops mainly doing repairs whenever the shoe business picked up or the garment industry slowed down. It was possible to make as much money as in garment assembly although the work was more intermittent; nonetheless, Juan Luis had more flexibility with his time, which offered other informal opportunities to earn money while also affording him the kind of mobility and independence that one had to sacrifice in the trade zones. When the local shoe business waned, he would return to the garment factory.

Juan Luis' livelihood strategy both affirmed and challenged the process of deskilling at the heart of the generalization of capitalist relations of production in Marx's classic account. The production of unskilled workers, a process that does not constitute a pre-history of capitalism but rather one dynamic of its on-going reproduction, is the very separation of workers from their powers of independent execution: that is, the realization of the Cartesian fantasy of disconnected bodies and minds, most often associated with the feminized labor of the global factory. "The possibility of an intelligent direction of production," Marx writes, "expands in one direction, because it vanishes in many others. What is lost by the specialized worker is concentrated in the capital which confronts them" (1976: 482). "Manufacture," Marx continues, "mutilates the worker, turning him into a fragment of himself [sic]" (1976: 482). Yet in investigating this process with relations of coloniality in mind, especially the centrality of the articulation of wage labor to complex geographies of work, we see that this classic story of proletarianization is open, contested, and never complete. In short, Juan Luis acceded to the forces that sought to render him a mere fragment in capitalist industrial production only ever temporarily, and drew upon his "whole," skilled body as a means to resist this process within the context of highly unstable wage and informal work. For the cabinet makers, carpenters, tailors, and shoemakers, like Juan Luis, who sold their labor to the trade zone factories when necessary and removed it when they could, the event of crisis that had rendered their skills "socially unnecessary" within the logics of capitalism was transformed into weekends, evenings, or longer hiatuses when they sought to reconstitute the value of their skills, and to subsidize their meager wages, within the informal economies of Santiago.[1] In short, crisis was transformed into a long-term practice of navigating the forces of capital that sought to devalue their labor and to deskill their work.

By far, the crisis that loomed largest in the narratives of *zoneros* was the country's financial crisis of 2003 under the government of Hipólito Mejía. The Dominican "miracle" of the mid- to late 1990s saw average annual growth rates of 5.7 percent (Sánchez Ancochea 2005). Mainstream economists celebrated the country's economic performance as a triumph of the 1990 market reforms, presenting the Dominican Republic as a model of success for the region (e.g., Wiarda 1999). The country's large, family-run corporate groups were the main beneficiaries of this apparent prosperity. Their fortunes grew, fueled in part by the loose lending practices of the banks that they controlled, lax oversight of those banks by government officials who had strong ties to these elites, as well as an unprecedented construction boom associated with land speculation in new tourist poles and in the capital city (Sánchez Ancochea 2005). In April of 2003, following eight months of government subsidies to avoid default, the country's second largest commercial bank, Baninter, faced collapse due to accumulated losses of 2.2 billion dollars. Despite evidence of massive fraud, and regulations that required the government to guarantee only up to 500,000 pesos per account and no obligation to guarantee offshore accounts (of which the bank had plenty), the Mejía government guaranteed all deposits. Few news stories reported that a total of 77 percent of the Baninter's deposits were held in *only* 85 accounts (Sánchez Ancochea 2005).[2] In effect, then, the bailout, representing about 13.5 percent of the country's GDP (UNCTAD 2009), constituted a massive transfer of public funds into private hands. As a result, public debt more than doubled, unemployment rose from 13 to 18 percent, and, due to soaring inflation, real wages declined by 22.5 percent in 2003 and again by 7.5 percent in 2004; meanwhile the currency depreciated from 28 to 52 pesos per dollar (Sánchez Ancochea 2005; World Bank 2006). Monica, a 12-year veteran of the trade zones in her late twenties, explained how she saw the financial crisis. "When the peso went up to 50, everything went up." "It was a massacre," she continued, "the factory owners took away bags of money and for the workers, we just got a few pesos raise." Earning in dollars and paying workers in now devalued currency, the crisis was a boon to trade zone owners and their multinational buyers who would benefit from lower factory prices. Many *zoneros* would repeat similar narratives to me in the context of their resentment towards an export class that had so poorly distributed the benefits of the country's restructured economy.

Ambrosina from Los Almácigos, whom we met in the book's opening vignette, had similar associations with the 2003 financial crisis. At the time of the crash, she and her husband, Joseph, had recently been laid off from an apparel factory that Baninter had embargoed several months prior due to the company's unpaid debts. Neither of them received severance pay and they had to move quickly to secure employment, ending up at IA Manufacturing. Ambrosina told me this story on a visit to her house in Ingenio Arriba, a working-class neighborhood near the Santiago trade zone built on the grounds of a shuttered sugar mill (called *un ingenio* in Spanish) from which the settlement took its name. To build her house in Santiago, Ambrosina had sold the agricultural land that had been given to her by her mother. "The land has no value," she commented in explaining her decision. With this money and her parents' help, Ambrosina had bought the wood of a small country house (*un rancho*) near her *campo* and moved the materials to Ingenio Arriba, where she and Joseph had built a small house on a side street populated by other migrants from Los Almácigos.

Despite her claim that her rural patrimony was value-less, Ambrosina was proud of her rural roots and closely connected to her *campo*. She conveyed this sentiment in her home by decorating the kitchen and small sitting room with old farm and household implements, such as a coal-fed iron that her mother used to use before the arrival of electricity to her hometown. She told me that she wanted her daughters to feel connected to the *campo* in their urban environment.

Two years after the Dominican financial crisis, in 2005, Ambrosina ran into a former supervisor from the factory that had been embargoed by Baninter. One-third of the bank's branches, together with its credit card and loan portfolios, had since been sold to the Canadian Bank of Nova Scotia. The former garment supervisor had found work refurbishing the Santiago Baninter branches for their new owner. Ambrosina arranged to get some materials from the renovation to extend her kitchen and to build a small back bedroom that she and Joseph could rent in order to earn more income. She acquired a television and a window, as well as a sink, a counter, and some cabinets from the branch's break room. The institutional off-white hue of the break room counter and cabinets in her kitchen stood in sharp contrast with the rustic slats of wood, painted pink and turquoise, of the rest of the house. In a literal sense, the detritus of the country's financial crisis had been reincorporated into the everyday spaces of reproduction. The recovered materials represented both a creative strategy and a necessity driven by the dynamics of spectacularly unequal wealth accumulation that the country's financialized economy had generated.

## Caribbean Crisis as Historical Debt

A powerful example of the simultaneously material and discursive force of crisis can be taken from contests over the framing of the notion of debt. Anthropologist David Graeber's expansive treatise on debt reminds us that relations of debt simultaneously entail (the threat of) violence, the mobilization of moral claims, and the reduction of social obligations to the cold calculus of monies owed. These elements work together. "If history shows anything," Graeber writes, "it is that there's no better way to justify relations founded on violence, to make such relations seem moral, than by reframing them in the language of debt" (2011: 5). The human consequences of debt – for example, the massive increases in poverty following structural adjustment in Latin America and the Caribbean, or the evictions of thousands of people from their homes in the wake of the US housing bubble – are made morally acceptable in a way that such outcomes stemming from bald theft simply would not. And yet, debt is rife with moral confusion, Graeber reminds us. For if paying one's debts has long been framed as a basic moral obligation, the usurer in almost all cultural contexts is, at one point or another, framed as morally bankrupt. Although debt, then, is generally wielded by the powerful to extract payment or tribute, it can also become a site of contestation or resistance framed through the very moral terms upon which the notion of debt depends. There is a fine and highly contestable line between socially acceptable debt and debt that becomes socially unsanctioned, pernicious, or, to use the term of the international debt forgiveness movement of the 1990s, odious. "Who *really* owes what to whom?" is a question that becomes possible precisely because of the moral foundations of debt that implicate debtor and creditor alike (Graeber 2011: 13; emphasis added).

While the attention of mainstream news agencies in the global North has been focused on the debt crisis of Southern Europe since 2008, much of the Caribbean has suffered from a similar fate. Since 2008, St Kitts and Nevis, Grenada, Belize, Antigua and Barbuda, and Jamaica have all faced default on their sovereign debt and have been forced to enter into IMF agreements to restructure their loans (IMF 2013). By 2013, Puerto Rico was also reported to be suffering from near insolvency, with public debt worth nearly 80 percent of the island's gross national product, forcing the government to issue municipal bonds at ever-higher interest rates merely to service its debt burden (Rodrigues & Bullock 2014). The case of Jamaica is particularly stark. In 2013, the country signed its second rescue package in three years for some 2 billion dollars and government debt remained at 143 percent above annual output. In addition to fiscal austerity, the Jamaican government agreed to allow its currency to depreciate in the hopes of stimulating exports and tourism, which caused crippling inflation and eroded the purchasing power of the working poor (IMF 2013). Mirroring the security discourses that I described in Haiti following that country's major consumption crisis of 2008, the contemporary debt crisis in the Caribbean has been mobilized to frame the entire region as the site of growing security threats from drug trafficking, enabled by high unemployment and weak state governance. In line with this discourse, the only major spending package that the United States has put forth in recent years in the region is the Caribbean Basin Security Initiative, a 263 million dollar initiative to train personnel and to upgrade the technological capabilities of states to interdict drug flows to the United States (US GAO 2013).

An exception to the paucity of attention surrounding the Caribbean's financial woes in the North Atlantic, English-language press is the *Financial Times*, which has dedicated numerous, lengthy articles that clearly reflect the narrative of international financial institutions on the current state of the region. One article quotes a high-level financial analyst and advisor to the Jamaican and Belizean governments on their debt restructuring who calls Caribbean sovereign debts "ticking time bombs" (Wigglesworth 2012). In fact, all of the financial experts together with the officials from the multilateral lending agencies quoted by the *Financial Times* express skepticism over the likelihood that the restructured loans will be repaid.[3] In addition, we read that some in the finance community have labeled Puerto Rico "America's Greece" (Rodrigues & Bullock 2014). Referring to the Anglo-Caribbean, an IMF official is quoted as saying that the region faces a "second lost decade" in reference to the 1980s when most Latin American and Caribbean economies countenanced a decade of zero or negative growth; and an official of the IDB calls the region's debt dilemma "an existential crisis" (Wigglesworth & Mander 2013).

For the finance community, the current debt crisis sweeping the region is the result of declines in tourism and remittances on the heels of recession in the North Atlantic and the paucity of other sources of foreign exchange. As we have seen in the cases of Haiti and the Dominican Republic, the promotion of exports like garment production has long relied upon "preferential" access to northern markets.[4] Other preferential arrangements included the postcolonial trade regime between the Anglo-Caribbean states and their former colonizers, which guaranteed markets for commodities like bananas produced in these Caribbean states by excluding the importation of bananas to Europe produced elsewhere.[5] All of these postcolonial and Cold War trade arrangements have been eroded in the face of growing competition from other world regions reinforced by the WTO trade regime, while few countries in the Caribbean have benefited from a substantial raw commodities boom that could finance their import bill. Thus, the "existential crisis" declared by the IDB official could be read as *a crisis of the creditor–debtor relationship itself*, for the very status of international lenders as creditors depends upon the earnings of their debtors to service their debts. Fundamentally, the production and trade arrangements at the heart of this debate, and this book, are also financial arrangements: if the countries of the Caribbean do not find their "niches" in the global economy to export or to earn tourist dollars, then they will cease to be viable clients of global finance capital, an unthinkable prospect for both lenders and borrowers alike.

The dilemma was captured powerfully by David Rockefeller, former head of Chase Manhattan Bank and patriarch of the Rockefeller family, who clearly framed the need for new export sectors in financial terms in the early 1990s on the heels of Latin American debt defaults of the 1980s. In the context of the debate on NAFTA with Mexico, which sought to bolster these new export sectors by locking in a supporting institutional framework (i.e., NAFTA), Rockefeller told an audience of influential industry leaders and politicians that "[this] route is the most efficient vehicle that we have to protect our own self-interest in maintaining the creditworthiness of Latin America" (1992, quoted in Fernández-Kelly & Massey 2007).

Rockefeller's statement reveals the moral ambiguity at the heart of debt relations and thus their vulnerability to moral challenges. Dominant crisis narratives precisely aim to regulate the conditions of narrative construction, to naturalize the market and thus to render unthinkable the interrogation of how precisely the creditor–debtor relationship is both produced and sustained. After all, who stands to gain from the creditworthiness of the Caribbean? It is only a short leap from this question to the one introduced as central in Graeber's analysis: who owes what to whom? Graeber warns that there is a danger in instigating a moral challenge along these lines since the question accepts the creditor's language even as it seeks to destabilize established power imbalances between creditor and debtor. Indeed, we should not presume that these are radical questions *per se*. Nonetheless, Caribbean leaders and intellectuals have continually seized upon these moral fissures in the foundations of transatlantic and north–south debt relations to attempt to wrest the crisis narrative from their former colonizers-now-creditors and from the United States. They, together with allies in Africa and other centers of the African diaspora, have long based their challenge on the damage and pernicious legacies of slavery, calling for reparations for this injustice.

The call for reparations for transatlantic slavery is as old as the abolition movement itself. The intellectual touchstone of twentieth-century efforts is Eric Williams' famous text *Capitalism and Slavery* (1994 [1944]). An early challenge to the teleology of development that posited individual countries "catching up" to industrial powers, especially Britain, Williams demonstrated the centrality of wealth accumulated through the slave trade to Britain's industrial revolution. Since the turn of the millennium, Caribbean scholars and political leaders have re-energized their efforts to subvert the dominant temporality of crisis of the kind espoused in the pages of the *Financial Times* by advancing their political claims for reparations. These claims took center stage at the historical World Conference against Racism, Discrimination and Xenophobia, held in Durban, South Africa in 2001. The fight to put reparations on the agenda for the first time at an official multilateral forum received fierce resistance from former colonial powers and from the United States (Beckles 2013). Shortly after, on the occasion of Haiti's bicentenary of independence, President Jean-Bertrand Aristide made the first-ever formal request by a Caribbean government for reparations from a former colonizer. Aristide requested the return of funds paid by Haiti to France under the threat of re-enslavement in the nineteenth century. Backed by a French blockade and battered by 20 years of political isolation, in 1825 Haiti had agreed to pay an exorbitant indemnity to the French government for lost property – that is, slaves and plantations – calculated to be roughly equivalent to 21 billion dollars today.[6] The country was forced to borrow money from France to make the first payment, inaugurating a cycle of debt that has never concluded.[7] In 2003, a beleaguered Aristide, facing a crippling aid embargo, called on France to repay the nineteenth-century indemnity. A French commission report on the matter, prepared by famed French socialist Régis Debray, denied the possibility of reparations while suggesting that aid served a similar function to reparations, an argument that would be repeated following the cancellation of Haiti's bilateral debt with France after the 2010 earthquake (Beckles 2013; Dubois 2013). The Haitian government abandoned the lawsuit following the removal of Aristide from power in 2004 (Beckles 2013).

The contestation over the framing of debt and the call for Caribbean reparations has been advanced with renewed impetus in the past few years. The timing is not accidental, as David Scott (2014) has recently argued, but rather responds to a particular postcolonial conjuncture of exhausted development models, on the one hand, and the rise of reparatory justice, on the other.[8] In the summer of 2013, the Caribbean Community (CARICOM), the region's economic integration organiza-tion, established the first-ever regional Reparations Commission for Caribbean Slavery and Native Genocide. The arguments advanced by the CARICOM Commission do not rest solely on the reversal of compensation to slave owners as demanded by Aristide, although such historical precedents lend powerful moral force to the claim. Rather, the Committee seeks "reparatory dialogue with beneficiary slave-owning European states with a view to formulating a new development agenda for the Caribbean [promoting] mutual respect, recognition, and concili-ation."[9] "Our constant search and struggle for development resources is linked directly to the historical inability of our nations to accumulate wealth from the efforts of our peoples during slavery and colonialism," explained Baldwin Spencer, the Prime Minister of Antigua and Barbuda, to the *New York Times* shortly after the Reparations Commission was established (Castle 2013). The Commission has identified six areas of Caribbean society that have suffered from the legacies of slavery and colonialism: public health, education, cultural institutions, cultural deprivation, psychological trauma, and scientific and technological "backwardness." All 15 member-governments have signed on to the Commission, which has hired a UK-based law firm to prepare a legal case potentially to be filed in the International Criminal Court. Thus, the reparations claim is not solely about mounting a moral challenge to the directionality of debt – who owes what to whom. Rather, the claim represents a more fundamental challenge aimed at unsettling the *aporia* of postcolonial history – that is, the denial of subject-hood and sovereignty. Reparatory justice, as sought by the Commission, appears as a "retemporalization of history; it attunes itself to a reenchanted past understood as a time not yet past that continues to disfigure the present and foreclose the future" (Scott 2014: ix). As a political challenge to the foundational unthinkability of slaves as historical agents and the legacy of silence in its wake, Caribbean leaders are advancing demands surrounding not solely monetary compensation, but also recognition and self-determination. In other words, these leaders are seeking not only to resignify debt, but also to imagine a fundamentally different notion of development.[10]

## Towards an Alternative Ethics of Global Production and Consumption

The dominant crisis frame authorizes the reproduction and entrenchment of established, unequal social relations: from the devaluation of labor by gales of "adjustment" to the transfers of funds from public coffers to private accounts and from the economically impoverished to the wealthy. Nonetheless, this discursive and material structure of power does not remain uncontested. Writers such as Kincaid and Trouillot construct accounts that create openings within the *aporia* of postcolonial history, openings that unsettle the dominant narrative. The everyday strategies mobilized by workers like Ambrosina and Juan Luis to marshal cultural and social resources to value their labor and livelihoods otherwise offer another vantage point from which to think about dislodging Eurocentric framings of crisis. Political campaigns to steer the debate on Caribbean debt towards the historic demand for reparatory justice mark a third. The three accounts of unsettling crisis that I offer the reader draw upon postcolonial and feminist politics to offer possible alternatives to redress the injustices of uneven development.

What insights might these critiques of crisis stemming from the Caribbean position within coloniality and capitalism lend to debates on ethics and the global factory? Today, the primary way to challenge exploitative relations of transnational production remains the mobilization of discourses and practices of "ethical consumption." Ethical consumption reinforces an understanding of global production as a relationship between consumers in the global North linked through webs of supply chains to producers in the global South. In the context of a neoliberal international trade regime, and confronted by the visible horrors of poorly regulated export industries in the global South (most recently in Bangladesh), consumers in the North have called for, and have helped to create, non-state mechanisms of social regulation for many globally produced and traded goods. Consumer-centered endeavors have sought fair trade labels and other kinds of certifications to ensure that the fruits of structurally devalued labor in the global South – from bananas, to coffee, to clothes and consumer electronics – are ethically produced. These efforts have led to modest gains when they respond to worker-led mobilizations, as we saw in the case of the CODEVI union in Chapter 5.

The ethical consumption framing, however, remains inadequate. Ethicality, in these endeavors, is equated with regular conditions of colonial capitalist exploitation, without challenging the premise upon which these unequal relations are made possible. While these ethical consumption efforts are important, if an ethics of global production begins and ends with ethical consumption, then such a politics merely reproduces, rather than transforms, existing relations of uneven development. Mimi Sheller makes a similar point in her sweeping history of Western consumption in/of the Caribbean, from its tropical products and biodiversity to the tourist gaze. These relations, no matter how "ethical" within the consumption frame, perpetuate asymmetries, "[implicating] some (northern) regions of the world in the material impoverishment of (tropical) others, and in forms of symbolic violence and cultural appropriation" (Sheller 2003: 4).

A critical engagement with the politics of global production must consider the coloniality of uneven development as "an active, structuring principal of present," to borrow from Stuart Hall (1980). Such an engagement cannot stop at the dominant framing of this relation as one between northern consumers and southern producers, an orientation that pervades both mainstream and many critical approaches to global production (Ramamurthy 2004). The proliferation of places and types of goods produced in the South for consumption in the North so central to our contemporary common sense about "globalization" offers an opportunity to think about these connections otherwise. These commodity links are in fact counter-flows of historical relations of coloniality that reproduce labor as "cheap" in an interminable cycle of repayment for an originary accumulation that continues to shape the value hierarchies of the global economy today. There are, of course, no easy solutions to these dilemmas but the voices brought together in this chapter and this book represent a modest contribution to the project of opening up horizons of thinking about North–South relations otherwise.

# Notes

1 Navigating between craftwork and the trade zones was a decidedly male strategy since apprenticeships and training in crafts and trades were rarely available to women. See my discussion in Chapter 3 on the incorporation of skilled, male tailors into the export garment industry.

2 For example, in its reporting on the collapse of Baninter and two smaller banks, the *Washington Post* emphasized the total number of depositors (around 400,000), "many of them individuals and small businesses," and failed to mention the distribution of funds among accounts (Wilson 2003).

3 See *Financial Times* articles by Robin Wigglesworth, particularly the in-depth series titled "Caribbean in Crisis," which ran in December 2013 and January 2014.

4 For an extended discussion of these arrangements between the United States and the circum-Caribbean, see Chapter 2.

5 Various iterations of these arrangements have been signed between the European Union and the African, Caribbean, and Pacific (ACP) group of countries since 1975. See "The Caribbean EPA Affair" by the late Caribbean economist Norman Girvan for a discussion of the neoliberal character of the treaties signed since the establishment of the WTO: http://www.normangirvan.info/wp-content/uploads/2009/08/pittsburgh-rev21.pdf. Accessed March 8, 2014.

6 Indeed, the amount extorted – 90 million gold francs – has been estimated to be equivalent to twice the country's "net worth" at the time (Beckles 2013). In parallel, in 1833, the British parliament spent 20 million pounds, or 40 percent of the country's annual expenditure, to compensate former slave owners in the wake of emancipation despite arguments by abolitionists in Parliament that former slaves, not their masters, should receive compensation. Scholars estimate the present-day value of the compensation to be around 11.6 billion pounds, or 18.6 billion dollars. Historians have traced the payments to the fortunes of numerous members of Britain's modern-day elites (Beckles 2013). These perverse indemnity payments – both Haiti's and the British government's – are drawn upon by reparations advocates as historical precedents and quantitative estimations of the wealth transfer to the perpetrators of slavery at the direct and indirect expense of their victims.

7 See Dubois (2012: 97–105), Nicholls (1996: 62–66), and Fatton (2011b). Indeed, the cycle of payment on this original indemnity to France did not conclude until 1881, by which time Haiti had accumulated other debts owing in part to the incredibly high burden of the original indemnity payments to France.

8 I am grateful to Matt Sparke for pointing me to this reference.

9 Press statement delivered by Professor Sir Hilary Beckles (Chairman) on behalf of the CARICOM Reparations Commission University of West Indies. December 10, 2013: http://caricom.org/jsp/pressreleases/press_releases_2013/pres285_13.jsp?null&prnf=1. Accessed March 8, 2014.

10  At the level of political discourse, this claim about development suggests, from my reading, the possibility for reparations to facilitate the basis of a welfarist state. Such an outcome would also of course depend upon national relations and social forces, in addition to the transformation of international relations of uneven development at the heart of the claim for reparations.

# 8

# Conclusion

The aim of this book has been to develop a set of conceptual tools to
engage with the complex geography of global production restructuring.
In light of the persistent discourse that represents the global factory as a
sequential stage of development, I have posited the global factory as an
unstable arrangement of power and difference that is both structured
and contingent. I have elaborated these tools and analytical insights
through a theoretically informed study of contemporary restructuring in
the Caribbean garment sector following multiple rounds of neoliberal
trade liberalization. My objective has been to elucidate how capitalist
relations of foreign investment and subcontracting depend upon and
reproduce geographies of uneven development over four decades.
Racialized and gendered geographies of uneven development serve as
axes of inequality upon which global production restructuring depends.
Uneven development is not only a reflection of capitalist processes of
dis/accumulation and structures of social difference, however. Rather,
uneven development is also shaped by the everyday struggles and
practices of workers and the unemployed who forge positions of social
worth under conditions not of their own choosing. Places are made
through this iterative process of capitalist accumulation, disinvestment,
and subaltern struggles. Here, I conclude by drawing out three implications of my account for our understanding of the relationship between
global production networks and geographies of uneven development in
the Caribbean and the global South.

*Global Displacements: The Making of Uneven Development in the Caribbean*, First Edition.
Marion Werner.
© 2016 John Wiley & Sons, Ltd. Published 2016 by John Wiley & Sons, Ltd.

# Labor In and Out of Global Production Networks

The study of the garment industry as a global production network offers key insights into the organizational, social, and geographical arrangements of transnational production. One motivation for this book has been to address a lacuna in our understanding of the role that labor plays in shaping these networks. I began my study by asking what we can learn about globalization from the threshold of an idled factory that once formed part of a transnational production network. To answer this question, I adopted a perspective that drew upon the notion of coloniality – or the colonial legacies that shape gendered and racialized hierarchies of labor today – and asked how those hierarchies intersect with arrangements of global production. Investigating the concrete determinations of these intersecting hierarchies of power (i.e., capitalist production networks) and difference (i.e., coloniality) in the garment sector in the Dominican Republic and Haiti, we can draw three conclusions with respect to labor.

First, global production network restructuring depends upon rearranging the functions of the labor process within and across firms and places. Hierarchies of social difference act as resources for this process of reproducing value by marshaling notions of natural skills and cultural traits in ways that are surprisingly malleable. So-called full package garment firms in Santiago that sought to insert themselves into more profitable niches of the production network, for example, did so through a slow process of excluding female sewing operators. These workers, who were hitherto deemed desirable for their socially constructed flexibility as unskilled and thus low-waged, were subsequently constructed as "too rigid" to adapt to the new demands of US buyers for smaller orders and more styles. The resulting sewing skill hierarchy in these garment firms was distinctly gendered, with self-trained, mostly male operators occupying the upper echelons, while their female counterparts occupied the lower rungs and faced more rapid retrenchment.

My argument is that capitalist accumulation proceeds by reworking both the functions in the labor process and the workers deemed appropriate for these tasks. A feminist perspective helps us to see in concrete terms that divisions of labor are not pre-determined and neither are the workers who fill these jobs (e.g., "unskilled," high-skilled). If feminist studies of global production have repeatedly made the latter claim, my emphasis has been more squarely on how this process takes place not only on the factory floor, but also through a sociospatial process shaped both by relations to global buyers and specific regional histories and cultures. In other words, workers are produced, not pre-given, as feminists have argued. And they are also not simply distributed within the hierarchical functions of global production and its lowest rungs of the "unskilled." The reworking of the functional division of labor in global production arrangements and the production of workers to fill these jobs shape one another and this process both depends upon – and refashions – geographies of uneven development.

The reproduction of skills and positions in global production networks is inseparable from the second process of restructuring that has been my focus here: the production of workers as unemployed, as non-value for capital. While worker disposability in the global factory has been an important theme in other studies (e.g., Wright 2006), the present text has focused on how the strategies of the un- and underemployed actively make places shaped by rounds of accumulation and disinvestment. Given the persistence of the export model in the Caribbean region over four decades based upon semi-mobile investment and contracting, I have argued that subaltern collective and individual strategies to forge positions of social worth may stem from capitalist restructuring but cannot be understood narrowly as responses to this process. Drawing upon the notion of geographies of work (Gidwani & Chari 2004; see Chapter 4), my argument has been that former garment workers' aspirational struggles constitute a skillful navigation of both precarious work – waged, informal, and unpaid – and state and social practices that seek to marginalize the bearers of these activities. These subaltern livelihood struggles – what I have called embodied negotiations – are inseparable from sociospatial divisions of labor that are gendered and raced in particular, historically specific ways. In the Cibao, for example, rural spaces were reproduced as places for Dominican male informal work and circumscribed as spaces of unpaid female labor. While these geographies of work make production network restructuring possible, they also far exceed the logics of this process. Decades of navigating the instabilities of production and periods of disinvestment have created a sociospatial terrain that simply cannot be understood through the singular lens of global production industries, no matter how dominant. Diverse livelihood strategies, and the social construction of identities in relation to them, shape the ways that workers submit to, navigate, and resist the capitalist wage relation. The articulation of racialized and gendered hierarchies not only with wage work, but also with a variety of additional livelihood strategies continually redraws the boundary between wage and non-wage work, and those included in, and excluded from, the former. What we can learn from the stories told here of the unemployed – of Marcos and his *motoconcho* and Yesenia and her rural food stand – is that the limits of capitalist production are horizons of subject-making that cannot be ignored.

Finally, if labor is significant as a constitutive outside to capitalist production as described here, the collective struggles of workers through unionization efforts are the most direct way in which labor shapes global industries. As we saw on the border, the formation of the CODEVI union challenged the hyper-exploitation of Haitian garment workers. Indeed, as of 2013, wages in the border trade zone were higher than those received by non-union garment workers in Port-au-Prince. Moreover, the presence of an independent union served as an example of worker empowerment and better labor conditions, one that garment employers struggled to suppress. Again, given dynamic restructuring and the transfer of production among factories within and between countries and macro-regions, the challenge to factory-based unions to "hold down" global production and to wrest a proportion of surplus for labor remains formidable.

## Regulatory Uneven Development

An analysis of the ebb and flow of the global factory in the Caribbean – and in Haiti and the Dominican Republic in particular – and the policy discourses that make sense of this process, demonstrate the ways in which neoliberal regulations are reproduced in the face of obvious failures. The status of the trade zone in development discourses and policies is a notable example of this process. As we saw in Chapter 3, the World Bank resignified trade zones in the Dominican Republic from spaces of market potential to signs of market burden. This international development policy discourse supported a shift in the national discourse, which projected a transition from "*manufactura*" to "*mentefactura*" (not dissimilar to projections of the "knowledge" economy in the North American context). On the surface, the World Bank's analysis of trade zones as regulatory burdens in the Dominican Republic contrasted strikingly to the celebration of trade zones by international financial institutions as the *sine qua non* of market-led reconstruction in Haiti that I discussed in Chapter 6. Yet, this apparent incongruity should be understood as one example of a kind of policy variegation that scholars have termed regulatory uneven development (Brenner et al. 2010). While the causes of this patterning are both historically specific and the result of local political struggles, the example points to the ways in which institutions in the global North reproduce their own power by adapting neoliberal logics to the variegated institutional terrain of the global South.

Regulatory uneven development, as we have seen, is also a process that depends in particular ways upon categories of social difference. We saw an example of this claim in the case of gendered imperatives both to employ women in trade zones as a form of market-driven poverty alleviation in the Dominican Republic in the 1980s and early 1990s and to justify the collapse of this model a decade later. In addition, development experts drew upon the colonial legacies and contemporary forms of anti-black racism in the initial proposal and eventual adoption of trade zone promotion as a form of threat containment in Haiti. These racialized and gendered discourses are drawn upon and reproduce the sort of patterning of regulatory uneven development that we have seen in the Caribbean. To make these connections usefully – between discourses and practices of social difference and forms of neoliberal restructuring – we must turn to grounded accounts of this process that interrogate the salience of social difference to particular processes of restructuring. As I have emphasized throughout this text, racism and patriarchy are obviously not un-changing structures and if we are to understand their reproduction in relation to capital accumulation, our accounts must continually ask the question of what difference social difference makes to capitalism.[1]

## Rethinking the Geographies of Uneven Development in the Caribbean and the Global South

The global factory, I have argued, is both a mechanism of capitalist accumulation and a set of assumptions, discourses, and spatial imaginaries that reproduce the notion of development as one of traversing a stage of industrialization. Mainstream development discourses have characterized the Dominican Republic as a successful adopter of the globalization model and Haiti as a failed entrant into global supply chains. These conclusions have circulated as common sense, validating the idea that the time has come for the Dominican Republic to "move on" from manufacturing to services, and for Haiti to "try again" to enter supply chains as a location for low-wage assembly. In contrast, the stories of worker livelihood strategies, union struggles, labor process restructuring, and development planning that I have brought together here demonstrate that global production arrangements are concrete determinations of social and spatial divisions of labor. Places do not "enter into" global supply chains; rather, places and global arrangements of production dynamically reproduce one another. If we displace the global factory as the primary entry point into our understanding of globalization, we face the task of grappling with the political and ethical dimensions of old and emerging patterns of uneven development.

The story of uneven regional development within and between these two countries offers the opportunity to consider the reproduction of long-standing divides into new geographies of uneven development in the global South. Geographers have long argued that peripheralization is a relational process that operates at multiple scales. The so-called "globalization" of production through new export manufacturing arrangements is in fact a mechanism to reproduce uneven development anew. Our task is to specify the particular geographies of uneven development formed through colonial legacies, capitalist relations, and the ways these forces articulate with regional conjunctures. Moving away from the study of the global factory to the very in/stabilities of global production, and the historically patterned and contingent geographies of uneven development that make this process possible, offers us a better model for understanding and engaging with new geographies of social and spatial inequality and injustice. The task of recuperating the notion of uneven development for the present, I believe, is an important one in order to better understand and thus to contest the iterative reproduction of colonial capitalism and hierarchies of social difference.

## Note

1   This formulation is indebted to Melissa Wright (personal communication) and to Spivak (1988).

# Bibliography

Abernathy, F. H., Volpe, A. & Weil, D. (2006) The future of the apparel and textile industries: Prospects and choices for public and private actors. *Environment and Planning A* 38 (12), 2207–2232.

Abrahamsen, R. (2004) The power of partnerships in global governance. *Third World Quarterly* 25 (8), 1453–1467.

Abreu, A. & Cocco, M. (1989) *Las Zonas Francas Industriales en la República Dominicana*. Centro Internacional para el Desarrollo Económico, Santo Domingo.

Amin, S. (1976) *Unequal Development: An Essay on the Social Formations of Peripheral Capitalism*. Monthly Review Press, New York and London.

Amsden, A. H. (1994) Why isn't the whole world experimenting with the East Asian model to develop? Review of "The East Asian miracle". *World Development* 22 (4), 627–633.

Amsden, A. H. (2003) Comment: Good-bye dependency theory, hello dependency theory. *Studies in Comparative International Development* 38 (1), 32–38.

Anglade, G. (1982) *Espace et Liberté en Haïti*. Université du Québec à Montreal et Centre de Recherches Caraïbes, Université de Montréal, Montréal.

Ariza, M. (2000) *Ya No Soy la que Dejé Atrás: Mujeres Migrantes en República Dominicana*. Editorial Plaza y Valdés, Mexico City.

Ariza, M. (2004) Obreras, sirvientas y prostitutas. Globalización, familia y mercados de trabajo en República Dominicana. *Estudios Sociológicos* XXII (64), 123–149.

Arrighi, G. (1994) *The Long Twentieth Century: Money, Power and the Origins of Our Times*. Verso, London.

Arrighi, G. & Drangel, J. (1986) The stratification of the world-economy: An exploration of the semiperipheral zone. *Review* X (1), 9–74.

*Global Displacements: The Making of Uneven Development in the Caribbean*, First Edition. Marion Werner.
© 2016 John Wiley & Sons, Ltd. Published 2016 by John Wiley & Sons, Ltd.

Arrighi, G., Silver, B. J. & Brewer, B. D. (2003) Industrial convergence and the persistence of the North–South divide. *Studies in Comparative International Development* 38 (1), 3–31.

Artiles-Gil, J. (2002) *Neoliberal discourse and the crisis of politics and culture: A comparative study of Dominican Republic and Costa Rica Grassroots politics during the eighties.* PhD Thesis, University of Minnesota, Minneapolis.

Augelli, J. (1962) Agricultural colonization in the Dominican Republic. *Economic Geography* 38 (1), 15–27.

Bair, J. (2005) Global capitalism and commodity chains: Looking back, going forward. *Competition & Change* 9 (2), 153–180.

Bair, J. (2009) Global commodity chains: Genealogy and review. In: Bair, J. (ed.) *Frontiers of Commodity Chain Research.* Stanford University Press, Stanford, CA, pp. 1–34.

Bair, J. (2010) On difference and capital: Gender and the globalization of production. *Signs: Journal of Women and Culture and Society* 36 (1), 203–226.

Bair, J. & Dussel Peters, E. (2006) Global commodity chains and endogenous growth: Export dynamism and development in Mexico and Honduras. *World Development* 34 (2), 203–221.

Bair, J. & Gereffi, G. (2001) Local clusters in global chains: The causes and consequences of export dynamism in Torreon's blue jean industry. *World Development* 29 (11), 1885–1903.

Bair, J. & Werner, M. (2011) Commodity chains and the uneven geographies of global capitalism: A disarticulations perspective. *Environment and Planning A* 43 (5), 988–997.

Bair, J. & Werner, M. (2015) Global production and uneven development: When bringing labour in isn't enough. In: Newsome, K., Taylor, P., Bair, J. & Rainnie, A. (eds.) *Putting Labour in its Place: Labour Process Analysis and Global Value Chains.* Palgrave Macmillan, Basingstoke, UK, pp. 119–134.

Bair, J., Berndt, C., Boeckler, M. & Werner, M. (2013) Dis/articulating producers markets and regions: New directions in critical studies of commodity chains. *Environment and Planning A* 45 (11), 2544–2552.

Balaguer, J. (1984) *La Isla al Revés: Haití y el Destino Dominicano.* Librería Dominicana, Santo Domingo.

Barrientos, S., Kabeer, N. & Hossain, N. (2004) The gender dimensions of the globalization of production. Working Paper No. 17. International Labour Office, Geneva.

Baud, M. (1995) *Peasants and Tobacco in the Dominican Republic.* University of Tennessee Press, Knoxville, TN.

Baud, M. (1997) Patriarchy and changing family strategies: Class and gender in the Dominican Republic. *The History of the Family* 2 (4), 355–377.

Beckles, H. (2013) *Britain's Black Debt: Reparations for Caribbean Slavery and Native Genocide.* University of West Indies Press, Kingston.

Bensusán, G. (2007) La efectividad de la legislación laboral en América Latina. No. 9290148403. Instituto Internacional de Estudios Laborales, OIT, Geneva.

Bergeron, S. (2003) The post-Washington Consensus and economic representations of women in development at the World Bank. *International Feminist Journal of Politics* 5 (3), 397–419.

Betances, E. (1995) *State and Society in the Dominican Republic*. Westview Press, Boulder, CO.

Blackwood, E. (2005) Wedding bell blues: Marriage, missing men, and matrifocal follies. *American Ethnologist* 32 (1), 3–19.

Brenner, N., Peck, J. & Theodore, N. J. K. (2010) Variegated neoliberalization: geographies, modalities, pathways. *Global Networks* 10 (2), 182–222.

Buck-Morss, S. (2000) Hegel and Haiti. *Critical Inquiry* 26, 821–865.

Bulmer-Thomas, V. (2003) *The Economic History of Latin America since Independence*. Cambridge University Press, New York.

Burawoy, M. (1979) *Manufacturing Consent: Changes in the Labor Process under Monopoly Capitalism*. University of Chicago Press, Chicago and London.

Burawoy, M. (1989) Two methods in search of science: Skocpol versus Trotsky. *Theory and Society* 18, 759–805.

Burawoy, M. (1998) The extended case method. *Sociological Theory* 16 (1), 4–33.

Burawoy, M., Blum, J. A., Sheba, G., et al. (2000) *Global Ethnography: Forces, Connections, and Imaginations in a Postmodern World*. University of California Press, Berkeley, CA.

Butler, J. (1993) *Bodies that Matter: On the Discursive Limits of "Sex."* Routledge, New York and London.

Cammack, P. (2004) What the World Bank means by poverty reduction, and why it matters. *New Political Economy* 9 (2), 189–211.

Candelario, G. (2007) *Black Behind the Ears: Dominican Racial Identity from Museums to Beauty Shops*. Duke University Press, Durham, NC and London.

Cardoso, F. H. & Faletto, E. (1979 [1971]) *Dependency and Development in Latin America*. University of California Press, Berkeley, CA.

Castle, S. (2013) 14 countries demand slavery reparations from former colonial powers; Caribbean nations hire legal firm to sue U.K., France and Netherlands. *The New York Times*, October 21, 2013, p. 6.

Castor, S. (1971) *La Ocupación Norteamericana de Haití y sus Consecuencias (1915–1934)*. Siglo Veintiuno, Mexico City.

Castro, M. & Boswell, T. (2002) The Dominican diaspora revisited: Dominicans and Dominican-Americans in a new century. *North-South Agenda Papers* 53, 1–25.

Chang, H.-J. (2006) *The East Asian Development Experience: The Miracle, the Crisis and the Future*. Zed Books, London and New York.

Chang, H.-J. (2007) *Bad Samaritans: The Myth of Free Trade and the Secret History of Capitalism*. Bloomsbury Publishing, New York.

Chari, S. (2004) *Fraternal Capital: Peasant-Workers, Self-Made Men, and Globalization in Provincial India*. Stanford University Press, Stanford, CA.

Charles, J. (2012) Promise of jobs spurs Haiti development debate. *Miami Herald* [Online]. June 10, 2012. Available: http://www.miamiherald.com/incoming/article1940498.html. Accessed February 2, 2015.

Childs, N. (2012) 2010/11 Rice Yearbook RCS-2012. US Department of Agriculture, Washington, DC.

Chomsky, N., Farmer, P. & Goodman, A. (eds.) (2004) *Getting Haiti Right This Time: The U.S. and the Coup*. Common Courage Press, Monroe, ME.

CIRH (Commission Intérimaire pour la Reconstruction d'Haïti) (2011) Projects approuvés par la CIRH: Mise à jour du statut. Bureau de performance et de lutte contre la corruption, Port-au-Prince.

Clifford, J. (1997) *Routes: Travel and Translation in the Late Twentieth Century*. Harvard University Press, Cambridge, MA.

CNC (Consejo Nacional de Competitividad) (2005) Plan nacional de competitividad sistémica de la República Dominicana. Presidencia de la República Dominicana, Santo Domingo.

CNZFE (Consejo Nacional de Zonas Francas de Exportación) (1999) Informe estadístico sector zonas francas, 1998. Santo Domingo.

CNZFE (Consejo Nacional de Zonas Francas de Exportación) (2005) Informe estadístico sector zonas francas, 1998. Santo Domingo.

CNZFE (Consejo Nacional de Zonas Francas de Exportación) (2009) Informe estadístico sector zonas francas, 1998. Santo Domingo.

CNZFE (Consejo Nacional de Zonas Francas de Exportación) (2013) Informe estadístico sector zonas francas, 1998. Santo Domingo.

Coe, N. M., Hess, M., Yeung, H. W., Dicken, P. & Henderson, J. (2004) 'Globalizing' regional development: A global production networks perspective. *Transactions of the Institute of British Geographers* 29 (4), 468–484.

Collier, P. (2007) *The Bottom Billion: Why the Poorest Countries Are Failing and What Can Be Done About It*. Oxford University Press, Oxford.

Collier, P. (2009) Haiti: From natural catastrophe to economic security (A report for the Secretary-General of the United Nations). United Nations.

Collins, J. (2003) *Threads: Gender, Labor and Power in the Global Apparel Industry*. University of Chicago Press, Chicago.

Cooper, F., Holt, T. C. & Scott, R. J. (2000) *Beyond Slavery: Explorations of Race, Labor, and Citizenship in Postemancipation Societies*. University of North Carolina Press, Chapel Hill, NC and London.

Corbridge, S. (1993) *Debt and Development*. Blackwell, Oxford, UK.

Coronil, F. (1996) Beyond Occidentalism: Toward nonimperial geohistorical categories. *Cultural Anthropology* 11 (1), 51–87.

Coughlin, D. & Ives, K. (2011) WikiLeaks Haiti: Let them live on $3 a day. *The Nation* [Online]. June 1, 2011. Available: http://www.thenation.com/article/161057/wikileaks-haiti-let-them-live-3-day. Accessed March 8, 2012.

Cox, R. W. (2008) Transnational capital, the US state and Latin American trade agreements. *Third World Quarterly* 29 (8), 1527–1544.

Craig, D. & Porter, D. (2003) Poverty reduction strategy papers: A new convergence. *World Development* 31 (1), 53–69.

Craig, D. & Porter, D. (2006) *Development Beyond Neoliberalism? Governance, Poverty Reduction, and Political Economy*. Routledge, New York.

Cravey, A. (1998) *Women and Work in Mexico's Maquiladoras*. Rowman & Littlefield, Lanham, MD.

Cravey, A. (2005) Working on the global assembly line. In: Nelson, L. & Seager, J. (eds.) *A Companion to Feminist Geography*. Blackwell, Malden, MA and Oxford, UK, pp. 109–122.

Crichlow, M. A. (2009) *Globalization and the Post-Creole Imagination: Notes on Fleeing the Plantation*. Duke University Press, Durham, NC.

Danticat, E. (1998) *The Farming of Bones: A Novel*. Soho Press, New York.

De Angelis, M. (2007) *The Beginning of History: Value Struggles and Global Capital*. Pluto Press, London.

De Córdoba, J. (2011) Planned Haitian textile park provides hope for jobs. *The Wall Street Journal* [Online]. January 11, 2011. Available: http://www.wsj. com/articles/SB10001424052748704458204576074152823637370. Accessed March 8, 2011.

Deere, C. (1990) *In the Shadows of the Sun: Caribbean Development Alternatives and US Policy*. Westview Press, Boulder, CO.

Demombynes, G., Holland, P. & León, G. (2010) Students and the market for schools in Haiti. Policy Research Working Paper Series WPS5503. World Bank, Washington, DC.

Derby, L. (1994) Haitians, magic and money: *Raza* and society in the Haitian–Dominican borderlands, 1900 to 1937. *Comparative Studies in Society and History* 36 (3), 488–526.

Derby, L. & Werner, M. (2013) The devil wears Dockers: Devil pacts, trade zones, and rural–urban ties in the Dominican Republic. *NWIG: New West Indian Guide* 87 (3&4), 294–321.

DeWind, J. & Kinley, D. H. (1988) *Aiding Migration: The Impact of International Development Assistance on Haiti*. Westview Press, Boulder, CO and London.

Diamond, J. (2005) *Collapse: How Societies Choose to Fail or Succeed*. Penguin, New York.

Dicken, P. (2011) *Global Shift, Sixth Edition: Mapping the Changing Contours of the World Economy*. Guilford Press, New York.

Dicken, P., Kelly, P., Olds, K. & Yeung, H. W. C. (2001) Chains and networks, territories and scales: Towards a relational framework for analysing the global economy. *Global Networks* 1 (2), 89–112.

Dilla, H. (2004) República Dominicana y Haití: Entre el peligro supuesto y el beneficio tangible. *Nueva Sociedad* 192, 23–34.

Dilla, H. & de Jesús, S. (2005) De problemas y oportunidades: Intermediación urbana fronteriza en República Dominicana. *Revista Mexicana de Sociología* 67 (1), 99–126.

Dore y Cabral, C. (1981) *Reforma Agraria y Luchas Sociales en la República Dominicana, 1966–1978*. Editora Taller, Santo Domingo.

Dubois, L. (2012) *Haiti: The Aftershocks of History*. Metropolitan Books, New York.

Dubois, L. (2013) OP-ED: Paying the price for Caribbean slavery. *The New York Times*, October 29, 2013, p. 29.

Dugan, I. J. & Dade, C. (2010) Medical care for Haitians falls short, group warns. *The Wall Street Journal* [Online]. January 22, 2010. Available: http://online.wsj.com/article/SB10001424052748704320104575015141368581502.html. Accessed November 1, 2011.

Dupuy, A. (1988) Conceptualizing the Duvalier dictatorship. *Latin American Perspectives* 15 (4), 105–114.

Dupuy, A. (2007) *The Prophet and Power*. Rowman & Littlefield, Lanham, MD.

Dupuy, A. (2010) Disaster capitalism to the rescue: The international community and Haiti after the earthquake. *NACLA Report on the Americas*.

ECLAC (Economic Commission for Latin America and the Caribbean) (2010) Economic survey of Latin America and the Caribbean: 2009–2010. Santiago, Chile.

Elie, J. R. (2006) *Participation, Décentralisation, Collectivités Territoriales en Haïti: La Problématique*. Bibliothèque Nationale d'Haïti, Port-au-Prince.

Elie, J. R. (2011) Quel intérêt pour la décentralisation, après le séisme du 12 janvier 2010? *Cahiers du CEPODE* 2 (2), 101–136.

Elson, D. (1979) The value theory of labour. In: Elson, D. (ed.) *Value: The Representation of Labour in Capitalism*. CSE, London, pp. 115–180.

Elson, D. & Pearson, R. (1981) 'Nimble fingers make cheap workers': An analysis of women's employment in Third World export manufacturing. *Feminist Review* 7, 87–107.

Escobar, A. (1995) *Encountering Development: The Making and Unmaking of the Third World*. Princeton University Press, Princeton, NJ.

Espinal, R. (1995) Economic restructuring, social protest and democratization in the Dominican Republic. *Latin American Perspectives* 22 (3), 63–79.

Evans, P. (1987) Class, state, and dependence in East Asia: Lessons for Latin Americanists. In: Deyo, F. C. (ed.) *The Political Economy of the New Asian Industrialism*. Cornell University Press, Ithaca, NY, pp. 203–226.

Evans, P. (1995) *Embedded Autonomy: States and Industrial Transformation*. Princeton University Press, Princeton, NJ.

Farmer, P. (1988) Blood, sweat, and baseballs: Haiti in the West Atlantic system. *Dialectical Anthropology* 13 (1), 83–99.

Farmer, P. (1994) *The Uses of Haiti*. Common Courage Press, Monroe, ME.

Farmer, P. (2008) Haiti's unnatural disaster. *The Nation* [Online]. October 6, 2008. Available: http://www.thenation.com/article/haitis-unnatural-disaster. Accessed October 12, 2012.

Farmer, P. (2011) *Haiti After the Earthquake*. PublicAffairs, New York.

Fatton, R. (2011a) Haiti's unending crisis of governance: Food, the constitution and the struggle for power. In: Heine, J. & Thompson, A. S. (eds.) *Fixing Haiti: MINUSTAH and Beyond*. United Nations University Press, Tokyo, pp. 41–65.

Fatton, R. (2011b) Haiti in the aftermath of the earthquake: The politics of catastrophe. *Journal of Black Studies* 42 (2), 158–185.

Federici, S. (2004) *Caliban and the Witch: Women, the Body and Primitive Accumulation*. Autonomedia, Brooklyn, NY.

Ferguson, J. (1999) *Expectations of Modernity*. University of California Press, Berkeley, CA.

Fernández, H. (2007) Sued: Cierre de zonas francas ha hecho proliferar negocios informales. *El Caribe*, August 14, 2007, p.1.

Fernández-Kelly, M. P. (1983) *For We Are Sold, I and My People: Women and Industry in Mexico's Frontier*. State University of New York Press, Albany, NY.

Fernández-Kelly, M. P. (1992) Labor force recomposition and industrial restructuring in electronics: Implications for free trade. *Hofstra Labor and Employment Law Journal* 10 (2), 623–700.

Fernández-Kelly, M. P. & Massey, D. S. (2007) Borders for whom? The role of NAFTA in Mexico–US migration. *The ANNALS of the American Academy of Political and Social Science* 610 (1), 98–118.

Fine, B. (2001) Neither the Washington nor the post-Washington Consensus: An introduction. In: Fine, B., Lapavitsas, C. & Pincus, J. (eds.) *Development Policy in the Twenty-First Century, Beyond the post-Washington Consensus.* Routledge, London, pp. 1–27.

Frank, A. G. (1967) *Capitalism and Underdevelopment in Latin America: Historical Studies of Chile and Brazil.* Monthly Review Press, New York.

Freeman, C. (2000) *High Tech and High Heels in the Global Economy.* Duke University Press, Durham, NC.

Freeman, C. & Murdock, D. (2001) Enduring traditions and new directions in feminist ethnography in the Caribbean and Latin America. *Feminist Studies* 27 (2), 423–459.

French, H. W. (1994) A Dominican's two burdens: Haiti and Balaguer. *The New York Times,* April 14, 1994, p. A4.

Fröbel, F., Heinrichs, J. & Kreye, O. (1980) *The New International Division of Labour: Structural Unemployment in Industrialised Countries and Industrialisation in Developing Countries.* Cambridge University Press, Cambridge, UK.

Georges, E. (1990) *The Making of a Transnational Community: Migration, Development and Cultural Change in the Dominican Republic.* Columbia University Press, New York.

Gereffi, G. (1994) The organization of buyer-driven global commodity chains: How U.S. retailers shape overseas production networks. In: Gereffi, G. & Korzeniewicz, M. (eds.) *Commodity Chains and Global Capitalism.* Praeger, Westport, CT, pp. 95–122.

Gereffi, G. (1999) International trade and industrial upgrading in the apparel commodity chain. *Journal of International Economics* 48 (1), 37–70.

Gereffi, G. & Korzeniewicz, M. (eds.) (1994) *Commodity Chains and Global Capitalism.* Greenwood Press, Westport, CT.

Gereffi, G. & Wyman, D. (1990) *Manufacturing Miracles: Paths of Industrialization in Latin America and East Asia.* Princeton University Press, Princeton, NJ.

Gereffi, G., Humphrey, J. & Sturgeon, T. J. (2005) The governance of global value chains. *Review of International Political Economy* 12 (1), 78–104.

Gertler, M. S. (1992) Flexibility revisited: Districts, nation-states and the forces of production. *Transactions of the Institute of British Geographers* 17 (3), 259–278.

Gibson-Graham, J. K. (1996) *The End of Capitalism (As We Knew It): A Feminist Critique of Political Economy.* Blackwell, Oxford.

Gidwani, V. & Chari, S. (2004) Guest editorial: Geographies of work. *Environment and Planning D* 22 (4), 475–484.

Gilmore, R. W. (2002) Fatal couplings of power and difference: Notes on racism and geography. *The Professional Geographer* 54 (1), 15–24.

Girvan, N. (2010) Technification, sweetification, treatyfication: Politics of the Caribbean–EU Economic Partnership Agreement. *Interventions* 12 (1), 100–111.

Glassman, J. (2004) Transnational hegemony and US labor foreign policy: Towards a Gramscian international labor geography. *Environment and Planning D* 22 (4), 573–594.

Glassman, J. (2011) The geo-political economy of global production networks. *Geography Compass* 5 (4), 154–164.

GOH (Government of Haiti) (2007) Growth and poverty reduction strategy paper (Document de Stratégie Nationale pour la Croissance et la Réduction de la Pauvreté, DSNCRP). Ministry of Planning and External Cooperation, Port-au-Prince.

GOH (Government of Haiti) (2009) Toward a new paradigm of cooperation. Conference on the Economic and Social Development of Haiti III. Washington, DC. April 14, 2009.

GOH (Government of Haiti) (2010) Action plan for national recovery and development: Immediate key initiatives for the future. Port-au-Prince.

González, R. (1993) Ideología del progreso y campesinado en el Siglo XIX. *Ecos* 1 (3), 26–43.

González de Peña, R. (2004) La figura del montero en la formación histórica del campesinado dominicano. *CLÍO: Órgano de la Academia Dominicana de la Historia* 74 (168), 75–96.

Graeber, D. (2011) *Debt: The First 5,000 Years*. Melville House, New York.

Grandin, G. (2006) *Empire's Workshop: Latin America, the United States and the Rise of New Imperialism*. Metropolitan, New York.

Grasmuck, S. (1982) Desarollo de enclave y el exceso relativo de la fuerza de trabajo: Mano de obra Haitiana en la República Dominicana. *Eme Eme: Estudios Dominicanos* X (60), 81–101.

Grasmuck, S. & Pessar, P. (1991) *Between Two Islands: Dominican International Migration*. University of California Press, Berkeley, CA.

Gregory, S. (2007) *The Devil Behind the Mirror: Globalization and Politics in the Dominican Republic*. University of California Press, Berkeley, CA.

Guarnizo, L. (1993) *One country in two: Dominican owned enterprises in New York and in the Dominican Republic*. PhD Thesis, Johns Hopkins University, Baltimore.

Gupta, A. & Ferguson, J. (1997) Discipline and practice: "The field" as site, method, and location in anthropology. In: Gupta, A. & Ferguson, J. (eds.) *Anthropological Locations: Boundaries and Grounds of a Field Science*. University of Chicago Press, Chicago, pp. 1–46.

Hadjimichalis, C. & Hudson, R. (2013) Contemporary crisis across Europe and the crisis of regional development theories. *Regional Studies* 48 (1), 208–218.

Hall, S. (1980) Race, articulation and societies structured in dominance. *Sociological Theories: Race and Colonialism*. UNESCO, Paris, pp. 305–345.

Hall, S. (1996) When was 'the post-colonial'? Thinking at the limit. In: Chambers, I. & Curti, L. (eds.) *The Post-Colonial Question*. Routledge, London and New York, pp. 242–260.

Hallward, P. (2007) *Damming the Flood: Haiti, Aristide, and the Politics of Containment*. Verso, London and New York.

Haraway, D. (1991) Situated knowledges: The science question in feminism and the privilege of partial perspective. In: *Simians, Cyborgs, and Women: The Reinvention of Nature*. Routledge, New York, pp. 183–201.

Hart, G. (1998) Multiple trajectories: A critique of industrial restructuring and the new institutionalism. *Antipode* 30 (4), 333–356.

Hart, G. (2002) *Disabling Globalization: Places of Power in Post-Apartheid South Africa*. University of California Press, Berkeley and Los Angeles.

Hartlyn, J. (1998) *The Struggle for Democratic Politics in the Dominican Republic*. University of North Carolina Press, Chapel Hill and London.

Harvey, D. (1989) *The Condition of Postmodernity*. Blackwell, Cambridge, MA and Oxford, UK.

Harvey, D. (1999 [1982]) *The Limits to Capital*. Verso, London.

Hein, S. (1992) Trade strategy and the dependency hypothesis: A comparison of policy, foreign investment, and economic growth in Latin America and East Asia. *Economic Development and Cultural Change* 40 (3), 495–521.

Heine, J. & Thompson, A. S. (eds.) (2011) *Fixing Haiti: MINUSTAH and Beyond*. United Nations University Press, Tokyo.

Henderson, J., Dicken, P., Hess, M., Coe, N. & Yeung, H. W. C. (2002) Global production networks and the analysis of economic development. *Review of International Political Economy* 9 (3), 436–464.

Heron, T. (2004) *The New Political Economy of United States-Caribbean Relations: The Apparel Industry and the Politics of NAFTA Parity*. Ashgate, Aldershot, UK.

Heron, T. (2006) An unraveling development strategy? Garment production in the Caribbean Basin after the Multi-Fibre Arrangement. *Bulletin of Latin American Research* 25 (2), 264–281.

Hoetink, H. (1980) El Cibao 1844–1900: Su aportación a la formación social de la República. *Eme Eme: Estudios Dominicanos* 8 (48), 3–19.

Hoetink, H. (1982) *The Dominican People, 1850–1900: Notes for a Historical Sociology*. Johns Hopkins University Press, Baltimore, MD.

Holmes, G. (2010) The rich, the powerful and the endangered: Conservation elites, networks and the Dominican Republic. *Antipode* 42 (3), 624–646.

Hopkins, T. K. & Wallerstein, I. (1977) Patterns of development of the modern world-system. *Review (Fernand Braudel Center)* 1 (2), 111–145.

Hopkins, T. K. & Wallerstein, I. (1986) Commodity chains in the world-economy prior to 1800. *Review* 10 (1), 157–170.

Hornbeck, J. F. (2010) The Haitian economy and the Hope Act. RL34687. Congressional Research Service, Washington, DC.

Howard, D. (2001) *Coloring the Nation: Race and Ethnicity in the Dominican Republic*. Signal Books, Oxford.

Howard, D. (2009) Urban violence, crime and threat to "Democratic Security" in the Dominican Republic. In: Mcgregor, D., Dodman, D. & Barker, D. (eds.) *Global Change and Caribbean Vulnerability: Environment, Economy and Society at Risk*. University of the West Indies Press, Kingston, pp. 298–316.

Hoy. (2004) Cardenal visita zona franca Juana Méndez. June 8, p. 4.

Hualde Alfaro, A. (2001) Trayectorias profesionales femeninas en mercados de trabajo masculinos: las ingenieras en la industria maquiladora. *Revista Mexicana de Sociología* 63 (2), 63–90.

Humphrey, J. & Schmitz, H. (2002) How does insertion in global value chains affect upgrading in industrial clusters? *Regional Studies* 36 (9), 1017–1027.

IDA & IFC (International Development Association & International Finance Corporation) (2011) Interim strategy note for the Republic of Haiti For CY 2012 Report No. 65112-HT. World Bank, Washington, DC.

IDB (Inter-American Development Bank) (n.d.) Haiti: Country strategy 2011–2015. Washington, DC.

IFC (International Finance Corporation) (2003a) Report to the Board of Directors on a proposed investment in [Dominican Textile], Dominican Republic and the Republic of Haiti. Project No. 20744. Washington, DC.

IFC (International Finance Corporation) (2003b) Summary of Project Information, Project 20744. Available: http://ifcext.ifc.org/ifcext/spiwebsite1.nsf/ProjectDisplay/SPI20744. Accessed June 7, 2014.

IFC (International Finance Corporation) (2005) Environmental Review Summary, Project 20744. Available: http://ifcext.ifc.org/ifcext/spiwebsite1.nsf/DocsByUNIDForPrint/F5FA909B2AC83174852576BA000E270B?opendocument. Accessed June 7, 2014.

IFC (International Finance Corporation) (2006) Press Release: IFC Adopts New Environmental and Social Standards. Available: http://www.csrwire.com/press_releases/21208--IFC-Adopts-New-Environmental-and-Social-Standards. Accessed June 7, 2014.

IMF (International Monetary Fund) (2013) Sovereign debt restructuring: Recent developments and implications for the fund's legal and policy framework. Washington, DC.

Isa Contreras, P., Ceara Hatton, M. & Cuello Camilo, F. (2003) *Desarrollo y Políticas Comerciales en la República Dominicana*. Friedrich Ebert Stiftung and CIECA, Santo Domingo.

Itzigsohn, J. (2000) *Developing Poverty: The State, Labor Market Deregulation, and the Informal Economy in Costa Rica and the Dominican Republic*. Pennsylvania State University, University Park, PA.

Jenson, J. (1989) 'The talents of women, the skills of men': Flexible specialisation and women. In: Wood, S. (ed.) *The Transformation of Work?* Unwin Hyman, London, pp. 141–155.

Kaplinsky, R. (1993) Export processing zones in the Dominican Republic: Transforming manufactures into commodities. *World Development* 21 (11), 1851–1865.

Kaplinsky, R. (2000) Spreading the gains from globalisation: What can be learned from value chain analysis? IDS Working Paper No. 110. IDS, Brighton, UK.

Katz, C. (1994) Playing the field: Questions of fieldwork in geography. *Professional Geographer* 46 (1), 67–72.

Kay, C. (1989) *Latin American Theories of Development and Underdevelopment*. Routledge, London.

Kelly, P. (2009) From global production networks to global reproduction networks: Households, migration and regional development in Cavite, Philippines. *Regional Studies* 43 (3), 449–461.

Kincaid, J. (1988) *A Small Place*. Farrar, Straus and Giroux, New York.

Klak, T. (2009) Development policy drift in Central America and the Caribbean. *Singapore Journal of Tropical Geography* 30 (1), 18–23.

Klein, N. (2007) *The Shock Doctrine: The Rise of Disaster Capitalism.* Alfred Knopf, Toronto.

Krippner, G. R. (2011) *Capitalizing on Crisis.* Harvard University Press, Cambridge, MA.

Kuczynski, P. & Williamson, J. (eds.) (2003) *After the Washington Consensus: Restarting Growth and Reform in Latin America.* Peterson Institute for International Economics, Washington, DC.

Lazarus, J. (2008) Participation in Poverty Reduction Strategy Papers: Reviewing the past, assessing the present and predicting the future. *Third World Quarterly* 29 (6), 1205–1221.

Lee, C. (1998) *Gender and the South China Miracle: Two Worlds of Factory Women.* University of California Press, Berkeley, CA.

Levien, M. (2013) Regimes of dispossession: From steel towns to special economic zones. *Development and Change* 44 (2), 381–407.

Lewis, A. (1949) Industrial development in Puerto Rico. *Caribbean Economic Review* 1 (1&2), 153–176.

Lewis, A. (1983 [1954]) Economic development with unlimited supplies of labor. In: Gersovitz, M. (ed.) *Selected Writings of W. Arthur Lewis.* New York University Press, New York, pp. 311–363.

Lipietz, A. (1986) New tendencies in the international division of labor: Regimes of accumulation and modes of regulation. In: Scott, A. J. & Storper, M. (eds.) *Production, Work, Territory: The Geographical Anatomy of Industrial Capitalism.* Allen & Unwin, London, pp. 16–40.

Lipietz, A. (1987) *Mirages and Miracles: The Crises of Global Fordism.* Verso, London.

Lopez-Acevedo, G. & Robertson, R. (2012) *Sewing Success?: Employment, Wages, and Poverty following the End of the Multi-Fibre Arrangement.* World Bank Publications, World Bank.

Lozano, W. (1993) Migración e informalidad en República Dominicana. *Ciencia y Sociedad* 18 (3), 254–270.

Lozano, W. (1998) *Jornaleros e Inmigrantes.* Facultad Latinoamericana de Ciencias Sociales: Instituto Tecnológico de Santo Domingo, Santo Domingo.

Lozano, W. (2001) *Los Trabajadores del Capitalismo Exportador. Mercado de Trabajo, Economía Exportadora y Sustitución de Importaciones en la República Dominicana: 1950–1980.* Banco Central de la República Dominicana, Santo Domingo.

Mañaná, V. (2004) Encapuchados tirotearon zona franca Juana Méndez. *Listín Diario,* February 16, 2004, p. A4.

Marcus, G. E. (1998) *Ethnography Through Thick and Thin.* Princeton University Press, Princeton, NJ.

Martin, P., Midgley, E. & Teitelbaum, M. S. (2002) Migration and development: Whither the Dominican Republic and Haiti? *International Migration Review* 36 (2), 570–592.

Marx, K. (1976) *Capital, Volume I*. Penguin Classics, London.

Massey, D. (1993) Power-geometry and a progressive sense of place. In: Bird, J., Curtis, B., Putnam, T., Robertson, G. & Tickner, L. (eds.) *Mapping the Futures*. Routledge, London, pp. 59–69.

Massey, D. (1995 [1984]) *Spatial Divisions of Labour: Social Structures and the Geography of Production*. Routledge, New York.

Matthews, D. T. (2002) Can the Dominican Republic's Export Processing Zones survive NAFTA? In: Gereffi, G., Spener, D. & Bair, J. (eds.) *Free Trade and Uneven Development: The North American Apparel Industry and NAFTA*. Temple University Press, Philadelphia, pp. 308–326.

McAfee, K. (1991) *Storm Signals: Structural Adjustment and Development Alternatives in the Caribbean*. Zed Books, London.

McKinley Jr, J. C. (2010) Homeless Haitians told not to try and flee to the US. *The New York Times*, January 18, 2010, p. A11.

McMichael, P. (2012) *Development and Social Change: A Global Perspective*. Sage, Los Angeles.

McMichael, P. (2013) Value-chain agriculture and debt relations: Contradictory outcomes. *Third World Quarterly* 34 (4), 671–690.

Meislin, R. J. (1984) Dominican aftermath. *The New York Times*, April 30, 1984, p. A9.

Menéndez, I. (1982) México al no alineamiento. *Nueva Sociedad* 63, 47–56.

Messina, W. A. & Seale, J. L. (1993) US sugar policy and the Caribbean Basin Economic Recovery Act: Conflicts between domestic and foreign policy objectives. *Review of Agricultural Economics* 15 (1), 167–180.

Mignolo, W. (1995) *The Darker Side of the Renaissance*. University of Michigan Press, Ann Arbor, MI.

Milberg, W. (2008) Shifting sources and uses of profits: Sustaining US financialization with global value chains. *Economy and Society* 37 (3), 420–451.

Mills, M. B. (2005) Engendering discourses of displacement: Contesting mobility and marginality in rural Thailand. *Ethnography* 6 (3), 385–419.

Mintz, S. (1974) *Caribbean Transformations*. Columbia University Press, New York.

Mintz, S. (1985) *Sweetness and Power: The Place of Sugar in Modern History*. Penguin, New York.

Mitchell, T. (2007) The properties of markets. In: Mackenzie, D., Muniesa, F. & Siu, L. (eds.) *Do Economists Make Markets? On the Performativity of Economics*. Princeton University Press, Princeton and Oxford, pp. 244–275.

Mohanty, C. T. (2003) *Feminism without Borders: Decolonizing Theory, Practicing Solidarity*. Duke University Press, Durham, NC.

Mortimore, M. (1999) Apparel Based Industrialization in the Caribbean Basin: A threadbare garment? *CEPAL Review* 67, 119–136.

Mortimore, M. & Peres, W. (1998) *Policy Competition for Foreign Direct Investment in the Caribbean Basin: Costa Rica, the Dominican Republic and Jamaica*. United Nations Economic Commission for Latin America and the Caribbean, Santiago, Chile.

Mortimore, M., Duthoo, H. & Guerrero, J. (1995) Informe sobre la competitividad internacional en las zonas francas. Naciones Unidas: Comisión Económica para América Latina y el Caribe, Santiago, Chile.

Moya Pons, F. (1986) *El pasado dominicano*. Fundación J. A. Caro Álvarez, Santo Domingo.

Moya Pons, F. (1990) Import-substitution industrialization politics in the Dominican Republic, 1925–1961. *The Hispanic American Historical Review* 70 (4), 539–577.

Moya Pons, F. (1992) *Empresarios en Conflicto*. Fondo Para el Avance de las Ciencias Sociales, Santo Domingo.

Mullings, B. (1999) Sides of the same coin?: Coping and resistance among Jamaican data-entry operators. *Annals of the Association of American Geographers* 89 (2), 290–311.

Mullings, B. (2005) Women Rule? Globalization and the feminization of managerial and professional workspaces in the Caribbean. *Gender, Place and Culture* 12 (1), 1–27.

Mullings, B., Werner, M. & Peake, L. (2010) Fear and loathing in Haiti: Race and the politics of humanitarian dispossession. *Acme: An international e-journal for critical geographies* 9 (3), 282–300.

Narotzky, S. & Smith, G. A. (2006) *Immediate Struggles: People, Power, and Place in Rural Spain*. University of California Press, Berkeley, CA.

Nathan and Associates (2009) Bringing HOPE to Haiti's apparel industry. CTMO-HOPE, World Bank, and Multidonor Trust Fund for Trade and Development.

*New York Times* (1984) Anatomy of a Caribbean riot. *The New York Times*, May 1, 1984, p. A30.

Ngai, P. (2005) *Made in China: Women Factory Workers in a Global Workplace*. Duke University Press, Durham, NC.

Nicholls, D. (1996) *From Dessalines to Duvalier: Race, Colour, and National Independence in Haiti*. Rutgers University Press, New Brunswick, NJ.

Ong, A. (1987) *Spirits of Resistance and Capitalist Discipline*. State University of New York Press, Albany, NY.

Ortiz, F. (1995 [1947]) *Cuban Counterpoint: Tobacco and Sugar*. Duke University Press, Durham, NC.

Padgett, T. (2010) Helping Haiti: The US Navy is ready but aid is not. *Time Magazine* [Online]. January 16, 2010. Available: http://content.time.com/time/specials/packages/article/0,28804,1953379_1953494_1954343,00.html. Accessed June 2, 2015.

Pantojas-García, E. (1985) The US Caribbean Basin Initiative and the Puerto Rican experience: Some parallels and lessons. *Latin American Perspectives* 12 (4), 105–128.

Pantojas-García, E. (1990) *Development Strategy as Ideology: Puerto Rico's Export Led Industrialization Experience*. L. Rienner, Boulder, CO.

Peck, J. & Tickell, A. (2002) Neoliberalizing space. *Antipode* 34 (3), 380–404.

Pérez Sáinz, J. (2003) Exclusión laboral en América Latina: Viejas y nuevas tendencias. *Sociología del trabajo* 47, 107–138.

Pinto Moreira, E. (2010) Till geography do us part? Prolegomena to an economic and monetary union between the Dominican Republic and Haiti. Policy Research Working Paper No. 5241. World Bank, Washington, DC.

Piore, M. & Sabel, C. (1984) *The Second Industrial Divide: Possibilities for Prosperity*. Basic Books, New York.

PNUD (Programa de las Naciones Unidas para el Desarrollo) (2005) Informe nacional de desarrollo humano: Hacia una inserción mundial incluyente y renovada. Santo Domingo.

PNUD (Programa de las Naciones Unidas para el Desarrollo) (2008) Informe sobre desarrollo humano: Desarrollo humano, una cuestión de poder. Santo Domingo.

Polanyi, K. (1957 [1944]) *The Great Transformation: The Political and Economic Origins of our Time*. Rinehart, New York.

Porter, D. & Craig, D. (2004) The Third Way and the Third World: Poverty reduction and social inclusion in the rise of 'inclusive' liberalism. *Review of International Political Economy* 11 (2), 387–423.

Portes, A., Dore-Cabral, C. & Landolt, P. (eds.) (1997) *The Urban Caribbean: Transition to the New Global Economy*. Johns Hopkins University Press, Baltimore, MD.

Portorreal, F. (1991) *Subordinación, resistencia y estrategias de lucha de las obreras textiles de la Zona Franca "San Isidro."* Departamento de Historia y Antropología, Universidad Autónoma de Santo Domingo, Santo Domingo.

Préval, R. (2010) President Barack Obama Holds a Media Availability with President Réne Préval of Haiti. CQ Transcriptions, Washington, DC.

Quijano, A. (1998) La colonialidad del poder y la experiencia cultural latino-americana. In: Briceño-León, R. & Sonntag, H. (eds.) *Pueblo, Época y Desarrollo: La Sociología de América Latina*. CENDES, LACSO, Nueva Sociedad, Caracas, pp. 27–38.

Quijano, A. (2000a) Coloniality of power, Eurocentrism, and Latin America. *Nepantla: Views from the South* 1 (3), 533–579.

Quijano, A. (2000b) Colonialidad del poder y clasificación social. *Journal of World-Systems Research* VI (2), 342–386.

Quiroga, L. (2003) Feminización de la matrícula universitaria en la República Dominicana: 1977–2002. Instituto Tecnológico de Santo Domingo, Centro de Estudios de Género, Santo Domingo.

Ramamurthy, P. (2004) Why is buying a 'Madras' cotton shirt a political act? A feminist commodity chain analysis. *Feminist Studies* 30 (3), 734–769.

Ramamurthy, P. (2010) Why are men doing floral sex work? Gender, cultural reproduction, and the feminization of agriculture. *Signs: Journal of Women in Culture and Society* 35 (2), 397–424.

Ramamurthy, P. (2011) Rearticulating caste: The global cottonseed commodity chain and the paradox of smallholder capitalism in south India. *Environment and Planning A* 43 (5), 1035–1056.

Ramírez, A. (2007) Empleados queman maquila: Descontentos, destruyen empresa al no recibir prestaciones. *Prensa Libre* [Online]. January 21, 2007. Available: http://www.prensalibre.com/pl/2007/enero/21/161398.html. Accessed January 30, 2007.

Ramírez, N., Santana, I., de Moya, F. & Tactuk, P. (1988) República Dominicana: Población y desarrollo. Comisión Económica para América Latina y el Caribe, San José.

Ravelo, S. & del Rosario, P. (1986) Impacto de los dominicanos ausentes en el financiamiento rural. Universidad Católica Madre y Maestra, Santo Domingo.

Raynolds, L. T. (1998) Harnessing women's work: Restructuring agricultural and industrial labour forces in the Dominican Republic. *Economic Geography* 74 (2), 149–169.

Raynolds, L. T. (2002) Wages for wives: Renegotiating gender and production relations in contract farming in the Dominican Republic. *World Development* 30 (5), 783–798.

Restrepo, E. & Rojas, A. (2010) *Inflexión Decolonial: Fuentes, Conceptos y Cuestionamientos*. Editorial Universidad del Cauca, Popayán, Colombia.

Reyes, R. (2001) El mercado de trabajo en República Dominicana, problemas y desafíos. Oficina Internacional del Trabajo, Santo Domingo.

Rich, A. (2001 [1985]) Notes towards a politics of location. In: *Arts of the Possible: Essays and Conversations*. W.W. Norton & Company, New York and London, pp. 62–82.

Rocheleau, D. & Ross, L. (1995) Trees as tools, trees as text: Struggles over resources in Zambrana-Chacuey, Dominican Republic. *Antipode* 27 (4), 407–428.

Rodrigues, V. & Bullock, N. (2014) Puerto Rico: Harbour of debt. *Financial Times* [Online]. January 23, 2014. Available: http://www.ft.com/intl/cms/s/0/cefd9c14-834c-11e3-86c9-00144feab7de.html#axzz3QDDMR4Gs. Accessed January 23, 2014.

Rodríguez Diaz, M. (1999) El Caribe en la estrategia norteamericana: El pensamiento geopolítico de Alfred T. Mahan. *Revista Mexicana del Caribe* 8, 66–88.

Rodrik, D. (2006) Goodbye Washington Consensus, Hello Washington Confusion? A review of the World Bank's economic growth in the 1990s. *Journal of Economic Literature* XLIV, 973–987.

Roitman, J. (2003) Unsanctioned wealth: Or, the productivity of debt in Northern Cameroon. *Public Culture* 15 (2), 211–237.

Roitman, J. (2014) *Anti-Crisis*. Duke University Press, Durham, NC and London.

Rosen, E. I. (2002) *Making Sweatshops: The Globalization of the US Apparel Industry*. University of California Press, Berkeley, CA.

Roy, A. (2010) *Poverty Capital: Microfinance and the Making of Development*. Routledge, New York and London.

Safa, H. (1990) Women and industrialisation in the Caribbean. In: Stichter, S. & Parpart, J. (eds.) *Women, Employment, and the Family in the International Division of Labor*. Macmillan, London, pp. 79–97.

Safa, H. (1995a) Economic restructuring and gender subordination. *Latin American Perspectives* 22 (2), 32–50.

Safa, H. (1995b) *The Myth of the Male Breadwinner*. Westview Press, Boulder, CO.

Safa, H. (2002) Questioning globalization: Gender and export processing in the Dominican Republic. *Journal of Developing Societies* 18 (2–3), 11–31.

Sagas, E. (1994) *An Apparent Contradiction? Popular Perceptions of Haiti and the Foreign Policy of the Dominican Republic*. Haitian Studies Association, Boston.

Saintilus, P. (2007) Ouanaminthe et la dynamique migratoire transfrontalière haïtiano-dominicaine. In: Calmont, A. & Audebert, C. (eds.) *Dynamiques Migratoires de la Caraïbe*. Karthala/Géode Caraïbe, Paris, pp. 323–339.

Salzinger, L. (2003) *Genders in Production: Making Workers in Mexico's Global Factories*. University of California Press, Berkeley, CA.

San Miguel, P. (1997) *Los Campesinos del Cibao: Economía de Mercado y Transformación Agraria en la República Dominicana 1880–1960*. Universidad de Puerto Rico, Puerto Rico.

San Miguel, P. (1999) *El Pasado Relegado: Estudios Sobre la Historia Agraria Dominicana*. La Trinitaria, Santo Domingo.

San Miguel, P. (2005) *The Imagined Island: History, Identity, and Utopia in Hispaniola*. University of North Carolina Press, Chapel Hill, NC.

Sánchez Ancochea, D. (2005) Domestic capital, civil servants and the state: Costa Rica and the Dominican Republic. *Journal of Latin American Studies* 37 (4), 693–727.

Sánchez-Fung, J. (2000) Empleo y mercados de trabajo en la República Dominicana: Una revisión de la literatura. *Revista de la CEPAL* 71, 163–175.

Santana, J. (1994) *Estrategia Neoliberal, Urbanización y Zonas Francas: El Caso de Santiago, República Dominicana*. FLACSO, Santo Domingo.

Santana, R. (2004) Suspenden labores en la zona franca de Juana Méndez. *Listín Diario*, June 10, 2004, p. A6.

Sassen, S. (1988) *The Mobility of Labor and Capital*. Cambridge University Press, Cambridge, UK.

Schrank, A. (2003) Foreign investors, "flying geese," and the limits to export-led industrialization in the Dominican Republic. *Theory and Society* 32 (4), 415–443.

Schrank, A. (2008) Export processing zones in the Dominican Republic: Schools or stopgaps? *World Development* 36 (8), 1381–1397.

Schumpeter, J. (2008 [1942]) *Capitalism, Socialism and Democracy*. Harper Perennial Modern Thought, New York.

Scott, D. (2014) Preface: Debt, redress. *Small Axe* 18 (1), vii–x.

Serra, N. & Stiglitz, J. E. (eds.) (2008) *The Washington Consensus Reconsidered: Towards a New Global Governance*. Oxford University Press, Oxford, UK.

Shamsie, Y. (2010) Time for a "High-Road" approach to EPZ development in Haiti. SSRC: Conflict Prevention and Peace Forum. SSRC, Waterloo, ON.

Shamsie, Y. (2011) Pro-poor economic development aid to Haiti: Unintended effects arising from the conflict–development nexus. *Journal of Peacebuilding & Development* 6 (3), 32–44.

Shamsie, Y. (2012) Haiti's post-earthquake transformation: What of agriculture and rural development? *Latin American Politics and Society* 54 (2), 133–152.

Sheller, M. (2003) *Consuming the Caribbean: From Arawaks to Zombies.* Routledge, London.

Sheppard, E. (2011) Geography, nature and the question of development. *Dialogues in Human Geography* 1 (1), 46–75.

Sheppard, E. & Nagar, R. (2004) From east–west to north–south. *Antipode* 36 (4), 557–563.

Silié, R. & Segura, C. (eds.) (2002) *Una Isla Para Dos.* FLACSO, Santo Domingo.

Silvey, R. (2000) Stigmatized spaces: Gender and mobility under crisis in South Sulawesi, Indonesia. *Gender, Place and Culture* 7 (2), 143–162.

Slater, D. (2004) *Geopolitics and the Post-colonial.* Blackwell, Malden, MA.

Smith, N. (2008 [1984]) *Uneven Development: Nature, Capital and the Production of Space.* University of Georgia Press, Athens, GA.

Sontag, D. (2012) Earthquake relief where Haiti wasn't broken. *The New York Times,* July 6, 2012, p. A1.

Sotelo Valencia, A. (2005) Dependencia y sistema mundial: ¿convergencia o divergencia? Contribución al debate sobre la teoría marxista de la dependencia en al siglo XXI. Mexico City.

Spivak, G. C. (1988) Scattered speculations on the question of value. In: *Other Worlds: Essays in Cultural Politics.* Routledge, New York and London, pp. 154–178.

Standing, G. (1999) Global feminization through flexible labor: A theme revisited. *World Development* 27 (3), 583–602.

Strittler, S. (2002) *In the Shadows of State and Capital: The United Fruit Company, Popular Struggle and Agrarian Restructuring in Ecuador, 1900–1995.* Duke University Press, Durham, NC.

Tewari, M. (2008) Varieties of global integration: Navigating institutional legacies and global networks in India's garment industry. *Competition and Change* 12 (1), 49–67.

Torres-Saillant, S. (1998) The tribulations of blackness: Stages in Dominican racial identity. *Latin American Perspectives* 25 (3), 126–146.

Traub-Werner, M. & Cravey, A. (2002) Spatiality, sweatshops and solidarity in Guatemala. *Social & Cultural Geography* 3 (4), 383–401.

Trotsky, L. (1969 [1906]) *The Permanent Revolution and Results and Prospects.* Merit Publishers, New York.

Trouillot, M. R. (1988) *Peasants and Capital: Dominica in the World Economy.* John Hopkins University Press, Baltimore, MD.

Trouillot, M. R. (1990) *Haiti: State Against Nation.* Monthly Review Press, New York.

Trouillot, M. R. (1992) The Caribbean region: An open frontier in anthropological theory. *Annual Review of Anthropology* 21, 19–42.

Trouillot, M. R. (1995) *Silencing the Past: Power and the Production of History.* Beacon Press, Boston.

Tsing, A. L. (2005) *Friction: An Ethnography of Global Connection.* Princeton University Press, Princeton, NJ and Oxford, UK.

Turits, R. (2002) A world destroyed, a nation imposed: The 1937 Haitian massacre in the Dominican Republic. *Hispanic American Historical Review* 82 (3), 585–630.

Turits, R. (2003) *Foundations of Despotism: Peasants, the Trujillo Regime, and Modernity in Dominican History*. Stanford University Press, Stanford, CA.

UN (United Nations) (2004) Republic of Haiti Interim cooperation framework 2004–2006: Summary report. United Nations, European Commission, World Bank, the Inter-American Development Bank.

UNCTAD (United Nations Conference on Trade and Development) (2009) Investment policy review: Dominican Republic. New York and Geneva.

Underhill, G. (1998) *Industrial Crisis and the Open Economy: Politics, Global Trade and the Textile Industry in the Advanced Economies*. Macmillan/St Martin's, London and New York.

US GAO (US General Accounting Office) (2013) Status of funding, equipment, and training for the Caribbean Basin Security Initiative. GAO-13-367R. Washington, DC.

USAID (US Agency for International Development) (2007) Dinámicas del desempleo en el sector textil de las zonas francas de la República Dominicana, entre el 2003 y el 2005. Greater Access to Trade Expansion, Office of Women in Development, Washington, DC.

USDA (US Department of Agriculture) (2000) Rice situation and outlook yearbook RCS-2000. Washington, DC.

Wade, R. H. (1996) Japan, the World Bank, and the art of paradigm maintenance: The East Asian miracle in political perspective. *New Left Review* 217, 3–37.

Wade, R. H. (2002) US hegemony and the World Bank: The fight over people and ideas. *Review of International Political Economy* 9 (2), 215–243.

Weisbrot, M. (1997) Structural adjustment in Haiti. *Monthly Review* 48 (8), 25–39.

Werner, M. (2010) Embodied Negotiations: identity, space and livelihood after trade zones in the Dominican Republic. *Gender, Place and Culture* 17 (6), 725–741.

Werner, M. (2011) Coloniality and the Contours of Global Production in the Dominican Republic and Haiti. *Antipode: A Radical Journal of Geography* 43 (5), 1573–1597.

Werner, M. (2012) Beyond Upgrading: Gendered labor and firm restructuring in the Dominican Republic. *Economic Geography* 88 (4), 403–422.

Werner, M. & Bair, J. (2009) After Sweatshops? Apparel politics in the Circum-Caribbean. *NACLA: Report on the Americas* 42 (4), 6–10.

Werner, M., Bair, J. & Fernández, V. R. (2014) Linking up to development? Global value chains and the making of a post-Washington Consensus. *Development and Change* 45 (6), 1219–1247.

Wiarda, H. (1999) Leading the World from the Caribbean: The Dominican Republic. *Hemisphere 2000* 7 (4), 1–4.

Wigglesworth, R. (2012) Finances of isle of St Kitts reflects woes of Caribbean. *Financial Times*, August 6, 2012, p. 8.

Wigglesworth, R. & Mander, B. (2013) A darkening debt storm; the Caribbean. *Financial Times*, April 29, 2013, p. 11.

Williams, E. (1994 [1944]) *Capitalism and Slavery*. University of North Carolina Press, Chapel Hill, NC.

Wilson, S. (2003) Dominicans suffer a reversal of fortune; Once thriving fiscal model turns sour. *The Washington Post*, November 30, 2003, p. A20.

Wolf, D. (1992) *Factory Daughters: Gender, Household Dynamics, and Rural Industrialization in Java*. University of California Press, Berkeley, CA.

World Bank (1978) Current economic position and prospects of Haiti. Report No. 2165–HA. World Bank, Washington, DC.

World Bank (1994) Export processing zones. Policy and Research. World Bank, Washington, DC.

World Bank (1995) The Dominican Republic, growth with equity: An agenda for reform. Latin American and Caribbean Regional Office, World Bank, Washington, DC.

World Bank (2006) Dominican Republic, country economic memorandum: The foundations of growth and competitiveness. Latin American and Caribbean Region, World Bank, Washington, DC.

Wright, M. (2001) Asian spies, American motors, and speculations on the space-time of value. *Environment and Planning A* 33 (12), 2175–2188.

Wright, M. (2006) *Disposable Women and Other Myths of Global Capitalism*. Routledge, New York and London.

Wynter, S. (2003) Unsettling the Coloniality of Being/Power/Truth/Freedom: Towards the Human, After Man, Its Overrepresentation – An Argument. *CR: The Centennial Review* 3 (3), 257–337.

Yunén, R. E. (1985) *La Isla Como Es: Hipótesis Para su Comprobación*. Universidad Católica Madre y Maestra, Santiago, República Dominicana.

# Index

value as social worth, 86–87, 98,
102, 183
value theory of labor, 81–82
value-added tariff, 34–35, 51 *see also*
outward processing trade
and backward linkages, 35

Wade, R., 50
wage rates
Dominican Republic, 140
Haiti, 129, 151
Haiti and Dominican Republic
compared, 60
Washington Consensus, 7, 35,
76–77, 143
Caribbean garment promotion, 35
currency devaluation, 35
labor regulation, 35
legitimacy crisis, 143
whiteness *see* hispanidad;
racialization
Williams, E., 24
Capitalism and Slavery (1944), 175
work
and disposability, 86
embodied negotiations, 87
geographies of, 86–87, 108
and post-emancipation societies, 87

unpaid domestic, 86
and wage/non-wage boundary, 85,
87, 95, 108
work hours, 2
worker resentment, 2, 96–97, 171
worker mobility *see also* gender
between factories, 2, 68, 100–101,
106, 169–170, 179
between trades and factory work,
67, 69, 170, 179
World Bank, 143, 161, 184 *see also*
International Finance
Corporation
critique of export assembly,
1970s, 151
and Dominican economy, 73–80
East Asian Miracle Report, 50
poverty reduction under James
Wolfensohn, 159
suspension of aid to Haiti, 122
world systems theory
commodity chain concept, 9
World Trade Organization
(WTO), 38
and trade preferences, 50, 174
and demise of postcolonial
arrangement, 174
Wright, M., 15, 56, 80, 86